12

FUELLING WAR:
Natural resources and armed conflict

PHILIPPE LE BILLON

ADELPHI PAPER 373

First published March 2005
by **Routledge**
4 Park Square, Milton Park, Abingdon, Oxon, OX14 4RN
for **The International Institute for Strategic Studies**
Arundel House, 13–15 Arundel Street, Temple Place, London, WC2R 3DX
www.iiss.org

Simultaneously published in the USA and Canada
by **Routledge**
270 Madison Ave., New York, NY 10016

Routledge is an imprint of the Taylor & Francis Group

© 2005 The International Institute for Strategic Studies

Director John Chipman
Editor Tim Huxley
Manager for Editorial Services Ayse Abdullah
Copy Editor Jill Dobson
Production Jesse Simon
Cover Images AP / Wide World Photos

Typeset by Techset Composition Ltd, Salisbury, Wiltshire
Printed and bound in Great Britain by Bell & Bain Ltd, Thornliebank, Glasgow

British Library Cataloguing in Publication Data
A catalogue record for this book is available from the British Library

Library of Congress Cataloguing in Publication Data

ISBN 0-415-37970-9
ISSN 0567-932X

CONTENTS

Maps, Tables and Figures

Glossary

ACP	African, Caribbean and Pacific countries	**MPLA**	Popular Movement for the Liberation of Angola
AFDL	Alliance of the Democratic Forces for the Liberation of Congo	**NGO**	Non-Governmental Organisation
BP	British Petroleum	**NPA**	New People's Army
CAISTAB	Coffee and Cacao Stabilisation Fund	**NPFL**	National Patriotic Front of Liberia
DRC	Democratic Republic of Congo (former Zaire)	**OCHA**	UN Office for the Coordination of Humanitarian Aid
ECOMOG	ECOWAS Monitoring Group	**OECD**	Organisation for Economic Cooperation and Development
ECOWAS	Economic Community of West African States		
EITI	Extractive Industries Transparency Initiative	**PNG**	Papua New Guinea
ELN	National Liberation Army	**RENAMO**	Mozambican National Resistance
FARC	Revolutionary Armed Forces of Colombia	**RTZ**	Rio Tinto-Zinc (now Rio Tinto plc.)
FLN	National Liberation Front	**RUF**	Revolutionary United Front
G8	Group of governments of eight major industrialized countries	**STABEX**	European compensatory financial scheme for ACP countries
GAM	Aceh Freedom Movement	**UNITA**	National Union for the Total Independence of Angola
GDP	Gross Domestic Product		
HRD	Diamond High Council	**UNSC**	United Nations Security Council
IMF	International Monetary Fund		
KPCS	Kimberley Process Certification Scheme	**UNTAC**	United Nations Transitional Authority in Cambodia
KR	Khmer Rouge (Party of Democratic Kampuchea)	**SPLA**	Sudan People's Liberation Army
MIF	Multinational Interception Force	**SWAPO**	South-West Africa People's Organization

Introduction

A generous endowment of natural resources such as oil, diamonds or timber should, in theory, favour a country's rapid economic and social development. Yet, from the oil fields of Iraq to the diamond mines of West Africa, millions of people in resource-rich countries have seen their lives devastated by the misuse of resource revenues. Economic dependence upon resource exports points to a paradox: compared to less well-endowed countries, resource-rich countries have been on average poorer and less competently governed. If not all resource-dependent countries follow this pattern, many have been governed by authoritarian and corrupt regimes, leaving the vast majority of their population under harsh conditions of poverty or blatant inequality. Long tolerated or supported by the rationale of Cold War geopolitics, many of these regimes have since fallen, for a variety of reasons: pressure for democratisation and respect for human rights, structural adjustments and the end of foreign patronage. In most cases, however, regime-collapse has aggravated the situation, with war occurring in many instances.

If not all resource-dependent countries are the same, neither are all resource sectors. Industrialised resource exploitation, such as offshore oil exploitation, facilitates the concentration of rents by governments and consolidates their rule. Much thus depends on the management of rents and their broader societal and geopolitical consequences. Other resource sectors, being highly susceptible to illicit exploitation and smuggling, easily resist fiscal control by governments. In such cases, alternative modes of resource control, such as the co-option of illicit economic actors, or sources of support for governments matter most.[1] The relative accessibility of resource revenues is an important factor in the balance of power between belligerents; following the decline of foreign patronage in a post-Cold War context, many belligerents have come to rely on resource revenues. In turn, this reliance has often led to an interpretation of hostilities as essentially driven by greed. Even United Nations peacekeepers, suggested Secretary General Kofi Annan, were not immune to the temptation of the 'poisonous mix' of diamonds and greed fuelling the war in Sierra Leone.[2]

Resources are not only linked to developmental failure and war financing. For political scientist Michael Klare, wars to control critical resources are becoming 'the most distinctive feature of the global security environment'.[3] Increasing demand for raw materials is driven by the spread of mass consumerism and sharpened global economic competition. Demand for oil, for example, is expected to grow at an average of about 2% per year in the next 20 years, but some pessimistic predictions suggest that maximum global oil production will be reached by as early as 2010. The instability of many producing countries, the expected oil shortage and the contested ownership of certain oil-producing areas also add up to tensions over resource supply. If geological constraints and technological advances can balance each other, political and economic factors remain major sources of concern for resource importers. In Klare's perspective, critical-resource scarcity will increasingly motivate military interventions as markets and technology fail to address perceived threats to supply.

According to these perspectives, resource-dependent countries should face a higher risk of war. Not all empirical studies confirm such higher risk exists, however, and the idea that resource dependence could cause war has been rejected from many quarters. As pointed out by many resource industries and human-rights advocates: it is people that kill people, not resources.[4] As put bluntly by an Angolan journalist jailed for denouncing corruption and war in his oil- and diamond-rich country: 'It's fashionable to say that we are cursed by our mineral riches. That's not true. We are cursed by our leaders'.[5] Available evidence suggests that conflict is not deterministically related to 'too much' or 'too few' resources. Rather, resource sectors may create a context in which a country's vulnerability to armed conflict is enhanced, and shape the opportunities available to would-be belligerents.

The core argument of this paper is that better governance of resource sectors can play a significant role in conflict prevention and termination. The case of Iraq clearly illustrates the dangers and costs of unaccountable regimes empowered by large resource rents. It also demonstrates the limitations and problems of foreign interventions in resource-rich countries, whether through economic sanctions or military invasion. While many developing countries have now diversified their economies and developed stronger democratic institutions, most African and Middle Eastern countries remain dependent on resource exports and are therefore the focus of this study. The remainder of the paper is structured around three questions: Why are resource-dependent countries particularly vulnerable to armed conflicts? How do resources influence armed conflicts? How can these problems be addressed?

The first chapter examines the idea of a 'resource curse' affecting many resource-dependent countries, according to which resource wealth tends to result in economic underperformance and governance failure.[6] Because of the volatility of international commodity prices and the opportunity for corruption offered by the size and discretionary control of most resource sectors, strong domestic institutions are required to cope with large economic shocks, and to allocate resource revenues fairly and efficiently. Yet, pre-existing state institutions in developing countries that are historically dependent on resources tend to be weak. This problem is often further aggravated by the dampening effect of large resource rents on political accountability: the availability of large rents controlled by the ruling elite reduces its fiscal reliance on broad-based taxation and thereby on the population. Accordingly, resource rents can help sustain the longevity of authoritarian and often corrupt regimes – as demonstrated by Saddam Hussein in Iraq or Mobutu Sese Seko in the former Zaire (now the Democratic Republic of Congo, DRC). Behind this façade of stability, the collapse of economic growth, governance failure and a trajectory of underdevelopment often lead to violent political transition processes. Besides the vulnerability resulting from economic and political factors, conflicts and violence within resource sectors themselves can set the stage for larger-scale hostilities, notably in relation to the claims and grievances of local minority groups in production areas.

Not all resources offer the same opportunities for belligerents. Chapter 2 identifies how different resources influence the course of armed conflicts, on account of their characteristics, location and mode of exploitation. Alluvial diamonds, for example, can sustain rebel groups for years when located in remote and difficult terrain, since they require minimal technology for extraction, are highly valuable and easily transportable, and until recently could be sold on the international market without significant controls. Direct exploitation of deep-shaft minerals or oil deposits is harder to achieve by rebel groups, since these are capital-intensive sectors more amenable to government control. These resources thus create a different opportunity context than do alluvial diamonds, and a coup d'état against the central state or a secession movement in the producing region may be the most effective strategies for rebel groups. Tactically, such opportunity context results in extortion schemes, theft, the kidnapping of staff and the targeting of vulnerable infrastructures.

Resource contexts can have a significant impact not only on the strategies and tactics adopted by belligerents, but also on the duration and impact of hostilities. Chapter 3 reviews recent attempts at curtailing the links between resources and armed conflicts, and discusses potential solutions to the resource curse and resource wars. It focuses in particular on business activities that largely insulate coercive governments or warlords from political accountability by providing essential taxation as well as investments and links to international markets. The context of weak public institutions and poor governance increases the need for corporate social responsibility. Resource businesses and primary commodity markets have a duty to foster better governance and address the needs of local populations. New international instruments, such as the Kimberley Process Certification Scheme for rough diamonds, or the Extractive Industries Transparency Initiative, demonstrate the importance of partnerships bringing together the private sector, advocacy groups and governments. Sanctions have been made more effective through better targeting and implementation. Reasserting the importance of resources on the security agenda, this paper ends with a call to drastically improve the governance framework of resource sectors.

CHAPTER ONE

The resource curse

In the aftermath of the Second World War and the rise of the development agenda, much hope was placed in the promise that their resource endowment would 'lift' many countries out of poverty. Not only would resource exploitation generate fiscal revenues and jobs, but also the necessary investment capital for an economic take-off. Windfall resource revenues, in other words, should prove a bonanza. Although Chile and Malaysia provide examples of countries that developed largely as a result of mobilising their resource sectors, such success stories are few and far between, and often tainted by authoritarian regimes and human-rights abuses. Rather, the experience of most resource-dependent countries appear to support the resource curse thesis.

Poor economic growth and exposure to shock.

Countries benefiting from a wealth of natural resources have experienced on average a lower economic growth rate than resource-poor ones over the past 30 years.[1] Small countries relying on mineral exports have been the worst affected, for example copper-exporting Zambia, which suffered greatly from the collapse of copper prices in the mid-1970s. Relative under-performance has also characterised many oil exporters.

Low standards of living, poverty and inequalities.

Mineral and oil dependence are correlated with lower levels of social development, such as child mortality.[2] Oil dependence is associated with high rates of child malnutrition, low health-care budgets, poor enrolment in primary and secondary education, and low adult literacy rates. High levels of mineral dependence are strongly correlated with higher poverty rates and lower life expectancy, and to a lesser extent, with greater income inequality. Even in the case of a 'success story' like that of Botswana, which benefited from sustained high levels of economic growth and significantly better governance than in most sub-Saharan African countries, and a per capita GDP of $6,872, prominent inequalities have left about 60% of the population of Botswana living on less than $2 a day.[3]

High levels of corruption.

Governments in resource-dependent countries tend to be more corrupt as a result of discretionary control over large resource rents.[4] Corruption appears to be particularly rife in less developed economies with weak institutions. Among the worst recent examples, the late General Sani Abacha reportedly embezzled an estimated $2.2bn over his four-year rule of oil-rich Nigeria.[5]

Authoritarianism and poor governance.

Both oil and mineral wealth appear to inhibit democracy and worsen the quality of governance.[6] Rather than promoting democratisation, resource rents tend to strengthen autocratic rule, all the more so in countries with poorly diversified economies. The predominance of autocracies among Middle Eastern oil producers is often used to illustrate this relationship. Although critics point to the specific history and culture of the Middle East, the correlation also holds true for resource-dependent countries in other regions.

Risk of civil war.

Economists Paul Collier and Anke Hoeffler have found that resource-dependent countries are more prone to civil war, the risk being highest when resource exports represent about a third of GDP. In their view, highly diversified economies are generally associated with industrialised democracies (a group largely insulated from civil wars), while highly resource-dependent countries have sufficient rents to buy social peace and buy off political opponents, and to build security agencies deterring large-scale rebellion (if not terrorism or foreign intervention). Powerful energy importers are also eager to ensure the stability of these countries – through military assistance and acquiescence to non-democratic rule and poor human-rights practices – although not always without backlash, as demonstrated by the case of Iran. More generally, abundant renewable resources in otherwise poor countries and non-renewable resources in all countries are reported to increase the likelihood of armed conflict.[8] While examples as diverse as Angola, Iraq and Papua New Guinea seem to confirm these findings, several studies have found that these relationships are not robust, with the possible exception of oil.[9]

The resource-curse thesis is not without its critics, however.[10] Most of the resource-curse symptoms characterise poor countries in general. Since many of these are resource-dependent, the resource curse could simply be a classic poverty trap. Yet negative economic effects resulting from resource windfall revenues have also affected industrialised countries. In the 1970s, the economy of the Netherlands was affected by what became known as the 'Dutch disease' as large foreign earnings from gas exports led an appreciation of the exchange rate that badly affected the manufacturing sector, an effect that has since been noted in the agricultural sectors of many developing countries benefiting from large mineral rents. Critics also point to the historical contingency of the resource curse. Although more research is needed, resource-rich countries were generally outdoing resource-poor ones economically until the 1960s. In the contemporary period, the oil shocks of 1973–78 and 1979–81 had severe consequences for most economies. Yet the impact on resource-dependent countries, including oil exporters, was more long lasting. Terms of trade declined for most primary commodity exports over the past 25 years, and most resource-dependent countries suffered from economic specialisation, the postponement of reform and debt overhang – not to mention the direct impact of outright rent embezzlement and depletion of resource reserves. Finally, critics point out that fuel or mineral-dependent economies have not under-performed on most economic and social indicators compared to other developing countries. As conceded by Richard Auty, economic geographer and strong proponent of the 'resource curse', the phenomenon 'is not an iron law, rather, it is a strong recurrent tendency'.[11]

No doubt the symptoms of the resource curse reflect broader causes and processes than those directly related to resource sectors. A country's political and economy history, its level of institutional development prior to resource discovery and exploitation, and the motivations and capacities of its leaders can all play a part. As such, resource-dependent countries are not equally exposed to the resource curse (and obviously not all wars stem from resource-related causes). Among the regions most negatively affected by resource dependence is Africa: a continent that relies on primary commodities for more than two-thirds of its exports; whose share in world exports declined from 6% to 2% between 1980 and 2002; and where the incidence of civil wars failed to decrease significantly over the 1990s, unlike most other regions in the world. The Middle East closely follows Africa, as most economies in that region have failed to diversify. Large oil rents have consolidated autocratic and often militaristic regimes, with Iraq under Saddam Hussein providing the most dramatic example.

Iraq and the resource curse

The oil sector played a growing role in Iraq during the twentieth century, from the colonial 'invention' of the country by British interests seeking to control potential oil-fields, to domestic political and economic life, and the decision of the US administration to invade Iraq in 2003 (Iraq's oil wealth made economic leverage over the regime of Saddam Hussein more difficult, and the war enabled a reorganisation of the Iraqi oil sector to make it more amenable to US interests). The responsibility of Saddam Hussein in the recent tragic history of Iraq cannot be denied, but beyond his individual responsibility lays a vast array of additional factors, one of which is oil wealth.

Economically, the Iraqi state benefited greatly from the 1970s oil boom, having nationalised its industry a few years before. Annual economic growth averaged 14% in the 1970s, and by 1979, Iraq was the second largest OPEC oil exporter behind Saudi Arabia. Oil represented about 55% of Iraq's GDP and more than 90% of government revenue. Oil revenues provided massive financial resources to the Iraqi state, strengthening the rule of its elite through patronage and coercion, and enabling a massive arms build-up that gave substance to increasing military ambitions. As historian Charles Tripp remarks, 'the restricted circles of the rulers and the primacy of military force have combined with the massive financial power granted to successive Iraqi governments by oil revenues to create dominant narratives marked by powerful, authoritarian leadership [and] the ideas of politics as discipline and of participation as conformity'.[12] If the Ba'ath Party allocated a vast proportion of this windfall to a massive arms build-up and the private interests of the regime's cronies, this windfall also benefited much of the population through quickly rising living standards.

Iraq's costly 1980–88 war with Iran and falling oil prices after 1981 resulted in a drastic economic turnaround. Negative growth averaged 6% per year during the 1980s. This growth collapse was further aggravated by the imposition of economic sanctions by the UN after Iraq's invasion of Kuwait in August 1990. By 1994, when the UN sanction regime was still in full force, Iraq's real per capita GDP was estimated to be close to that of the 1940s.[13] This situation was improved under the UN Oil-For-Food Programme, but by the time of the 2003 US-led invasion, Iraq faced staggering financial obligations estimated at up to $383 billion, a devastated economic and public service infrastructure in need of an estimated $53bn and a largely destitute population.[14]

If oil allowed the Ba'athist party-state to extend the reach of its patronage and security apparatus, the regime proved relatively ineffective at building consensual national politics and a strong economy. As Iraqi academic Isam al Khafaji remarks, with the 1970s oil windfall the regime,

> forcibly imposed a destructive concept of unity among Iraqis which sought, and succeeded to a certain extent to atomise the population and linking the individuals directly to the state ... marginalising and suppressing entire communities and regions ... political, family and clannish cronyism deprived Iraqis from any autonomy and enhanced a perception among them that the state does not owe anything to the people. Rather it was they who owed their living to the state. [15]

As the state withdrew from many basic social and economic services from the mid-1980s onwards, non-state institutions such as religious organisations once again asserted their social importance while the regime continued to rely on large-scale domestic political violence to ensure its survival. In the context of 13 years of UN sanctions, corruption and arbitrary rule became central elements in the survival of the Iraqi regime: the entourage of Saddam Hussein wielded tight control over smuggling and clamped down on competing sanction-busters.[16] Illicit earnings from oil smuggling and kickbacks under the Oil-for-Food Programme between 1997 and 2002 amounted to an estimated $10.1bn.[17] Although the concentration of wealth within the ruling family is certainly not exceptional in the region, the level of neglect and abuse of the population certainly was.

The current Iraqi state will not be in a position to regain such hegemony in Iraqi society in the next few years. With annual petroleum revenues forecasted at only around $20bn dollars in the coming years – compared to about $50bn at the height of the 1970s for a population half its current size – the fiscal leverage of the fledgling post-Saddam state remains limited; all the more so as insurgents and looters continue to incapacitate the oil sector.[18] Ironically, this situation may help to ensure political plurality, but it also leaves it more depend on foreign donors and their agendas. In the context of the insurrection against the US presence and the internecine conflicts between rival Iraqi groups, a financially weak and donor dependent Iraqi state is a risky proposition. Oil revenue shortfalls will limit the capacity of the state and legitimacy. Ultimately, the governance of the oil sector will have a major role in the success or failure of the current political and economic transition in Iraq.

The tragic history of Iraq is not unique, as demonstrated by the plight of populations in many African oil-dependent countries.[19] Iraq's path of underdevelopment, however, should not be over-generalised. Sound economic policies and institutions can adapt to and tackle the many challenges of resource dependence. Yet, as political scientists Michael Shafer and Terry Karl have demonstrated, resource-dependent states often embody extreme levels of institutional conservatism and inertia, as organised interests and state bureaucrats controlling resource rents fight to maintain a status quo in their favour, while governments rely on fiscal transfer rather than statecraft to sustain the regime.[20] In short, resource dependence does not condemn a country to developmental failure, and even less so to war, but nevertheless represents a potential source of problems, creating conditions conducive to the outbreak of war. In this regard, attention should be paid to the negative economic and political impact of resource dependency and to the conflicts and various forms of violence directly associated with resource exploitation.

The collapse of economic growth

Resource exploitation can provide a valuable source of foreign exchange, employment and technological transfer. Yet, the general picture is one of relative economic underperformance in resource-dependent countries, with most empirical studies demonstrating that such dependence can lead to the collapse of economic growth, a major factor of instability and armed conflicts. Resource dependence can adversely affect the economy through a number of factors, including declining terms of trade and revenue shocks, budgetary mismanagement and negative effects on non-resource economic sectors, which are poorly integrated with the resource sector and rarely benefit from resource wealth.

Since the 1970s, the decline and volatility of most primary commodity prices have significantly reduced per-capita economic growth in exporting countries.[21] Non-fuel primary commodity prices have declined on average by about 30% over the past 30 years. Not only are declining prices affecting terms of trade, but price volatility is making budgetary and long-term planning major challenges. The price of oil, for example, increased nearly sixfold between December 1998 and March 2005. Between 1957 and 1999, primary commodity prices on average experienced a boom-and-bust cycle every six years, with prices falling 46% during slumps and rising 42% during booms.[22] Responses to declining prices have varied. But while government economic policy and the quality of governance and institutions are certainly relevant to the resource curse, these appear to have made little difference to the generalised economic slowdown of many resource-dependent countries over the last 30 years.

Low resource prices in the context of increased international competition have promoted a cost-cutting and consolidation strategy affecting the least productive and riskiest projects. In the context of structural adjustments, and with more capital and technology deployed to access resources and increase productivity, local governments have lost out to multinational corporations through massive privatisations and in negotiations over revenue sharing. Political risks and the foreign investment climate have also become more important. Many multinationals prefer to face higher exploitation costs and technological challenges than political risks that could affect their reputation or long-term investments. The threat of closure for less profitable resource projects has sometimes pushed local governments to subsidise their resource industries, but at the cost of greater fiscal deficit and further depression of international commodity prices. The massive subsidisation of their own resource sectors by industrialised countries has been internationally condemned for its effect on poor resource-exporting countries; for example, US subsidies and tariffs in the cotton and steel sector.

In times of high resource prices, over-extraction of rents from the resource sector has been common, preventing investment in maintenance or in the development of new projects. In many places, small-scale activities concentrating on the most accessible and profitable deposits have replaced large industrialised projects that fell prey to predatory practices, international competition and prolonged price slumps. Such fragmentation and scaling-down of resource exploitation in turn make it more difficult for governments to tax the resource sector and to ensure revenues essential to the maintenance of basic infrastructure and services. As a result, many African countries have lost their position on international commodity markets, with diamonds, gold and oil the few remaining sectors that thrive.

The effects of price volatility generally go well beyond the immediate boom-and-bust cycles, as resource-dependent countries tend to borrow money rather than drastically cut budgets during slumps. Not only do prices fall more on average than they rise during the cycle, but slumps also tend to last longer than booms. Moreover, many governments tend to discount the fact that revenues rely upon finite resource reserves, hoping that new investments and technologies will discover new reserves. Short-term borrowing often turns into costly debt overhang when prices fail to recover, no new commercial reserves are found, or after loans are misman- aged or squandered in corruption and 'white elephant' projects (such as the lavish hosting of the Organisation of the African Union by the Sierra Leone government in the midst of a fiscal crisis in the late 1980s).[23] Even without corruption, clientelist politics led many governments to eschew coherent economic policies that maximise long-term social welfare for the short-term management of political constituency demands and social tensions. This trade-off results in inefficient investment and low growth. If the resource rent proves insufficient to dampen demands for reform, the state becomes more vulnerable and social tensions increase, while the opportunity cost of dissent decreases.

Many examples can be drawn from both the Middle East and Africa. After a decade of rising oil prices, the fall in prices from $52 to $15 per barrel in the mid-1980s debilitated many oil-producing economies and jeopardised their political systems.[24] In Saudi Arabia, economic growth has not kept pace with the expanding population and its aspirations. This has resulted in a significant decline in per-capita income, growing inequal- ities between the extended royal family and the general population, and support for Islamic fundamentalism and terrorism. In Algeria, the Front Islamique du Salut increased its popular support by denouncing the fail- ure of the Algerian government's petroleum-driven economic model of state-led industrialisation and a 'socialist market', and by exploiting the economic and social difficulties of the regime following its mismanage- ment of the post-1986 fiscal crisis when oil prices slumped.[25]

A potentially prosperous oil-producing country, the Republic of Congo (Brazzaville) is one of Africa's most highly indebted countries on a per- capita basis.[26] When oil prices collapsed in the mid-1980s, the govern- ment turned to oil-collaterised loans to avoid restructuring the economy or downsizing its bloated bureaucracy. As oil prices failed to recover, and in the absence of an alternative economic base, debt servicing increased and the government was unable to pay civil servants. This growth collapse provided the context in which a fraught democratisation led to civil war in the mid 1990s. Under the 17-year rule of Siaka Stevens, a similar process of declining mineral revenues, corruption and indebtedness weakened the state in Sierra Leone in the 1980s. Having lost both official and informal control over a diamond sector that was key to political patronage and state revenues, Stevens' political heir, Joseph Momoh, was unable to suppress an insurrection supported by neighbouring Liberian warlord Charles Taylor in the early 1990s.[27]

In Rwanda, the dependence of segments of the ruling elite and many farmers on coffee exports was a structural factor in the country's political and economic vulnerability. The poor performance of the International Coffee Agreement regulating world coffee prices led to a drastic fall in prices from the mid-1980s onwards, and the agreement's ultimate collapse in 1989. Along with the collapse of the mining sector, the decline of coffee export earnings from $144m in 1985 to $30m in 1993 strained state finances, increased hardship among farmers and hurt the private interests of many in the ruling elite (the massacre in Burundi of tens of thousands of Hutus by the Tutsi-dominated army in 1972 after a long period of stable coffee prices, however, demonstrates the limits of this argument).[28]

Economic diversification should offer a simple solution to the economic impacts of the resource curse. Yet many resource-dependent countries have failed to diversify their economies. One reason is the lack of productive economic linkages between the resource sector and the rest of the economy. This is particularly the case for oil and deep-shaft mineral sectors, which are capital-intensive and rarely generate large direct employment. This absence of productive linkage is amplified when resource revenues pay for imported rather than locally produced consumer goods (in particular, imported food which depresses the local agricultural sector), or pay for non-productive employment in state bureaucracies. Attempts to use resource revenues to support non-resource sectors – for example, through the protection of 'infant industries' – often fail to achieve real competitiveness, especially when projects are waylaid by corruption.

Among other factors impeding diversification, the 'Dutch disease' mentioned above can negatively affect non-resource sectors by reducing the competitiveness of local goods and services on domestic and international markets through the appreciation of the local currency, while rising demand in domestic service sectors may lead to inflation. In non-diversified resource-rich economies, occupational specialisation is low and socio-cultural changes slow, which in turn affect the growth potential of the rest of the economy. Rent-seeking government officials and elites also tend to allocate efforts and investments to the resource sector rather than other activities.[29] Moreover, rulers benefiting from resource rents have little incentive to diversify an economy that could foster political competition. Similarly, privatisation is geared towards benefiting cronies; and keeping them out of politics, as illustrated by the arrest of the former CEO of Russian oil major Yukos, Mikhail Khodorkovsky. Rulers can further ward off such risk of competition from successful domestic business actors 'meddling' in politics by devolving the management of resource sectors

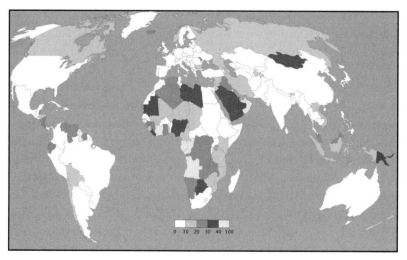

Map 1 – *Primary commodity export dependence* (% OF GDP, 1995)

to foreign firms (or domestic yet 'foreign' minority groups, as is largely the case with 'Lebanese' diamond exporters in West Africa). This strategy also offers the advantage of satisfying international financial institutions and consolidating external political support through 'private' diplomacy driven by business interests. In this way, resource businesses and their home governments can become a key, generally docile 'political constituency' of the regime, indirectly serving to further marginalise the voice of the local population. Besides economic factors, governance is a key dimension of the resource course and its impact on armed conflicts.

Governance failure

Not only can resource dependence negatively affect economic performance, but also politics and the quality of governance. Given the opportunity, ruling groups and communities routinely manipulate economic sectors for political and financial gains. Resource-based sectors lend themselves to political control while increasing the rewards of ruling, because natural resources generate large rents falling in state hands through parastatal companies or taxation. Ruling groups can in turn take advantage of available resource rents to consolidate their power independently of the local population and better resist international pressure. To do so, many ruling groups systematically undermine the very institutions that could mitigate these perverse political incentives. Competing state institutions, such as resource management departments or an independent judiciary, fall prey to a pattern that also extends into political patronage, intimidation, legislative interference, or outright kleptocracy. The consequences of this institutional weakening often reach deep into the rest of the economy and society. In many ways, the population finds itself 'hostage' to a state that is both a source of hope and frustration, but which faces limited 'incentives to bargain with [its] own citizens over resources or to institute or respect democratic processes around public revenue and expenditure'.[30]

Several dampening effects of resource dependence on democracy and the quality of governance have been identified:

- *rentier-state effect*, whereby the availability of resource rents to the government allows it to pursue the politics of patronage towards the population, thereby undermining democratic representation and the accountability achieved through the fiscal bargaining and institution-building necessitated by broad-based taxation in diversified economies;
- *repression effect*, occurring when resource rents financially support higher internal security expenditures, warding off democratic pressure;
- *modernisation-retarding effect*, resulting as economic growth based on raw materials fails to bring about socio-professional and cultural changes that promote democracy and a thriving civil society;
- *foreign political interference effect* resulting from the resource interests of foreign powers, whereby autocratic rulers are imposed or supported, or forced democratisation precipitates a political crisis.

Political scientist Michael Ross finds tentative support for the first three of these causal elements in the case of oil exporters, and for the rentier effect and to a lesser extent, for the modernisation-retarding effects in the case of non-fuel mineral exporters.[31] Population movements, such as rapid urbanisation and migrant labour, and the rise of a cash economy that characterises many resource booms, can accelerate social change and demands for political representation. Although anecdotal evidence can be drawn from historical cases like the former Zaire, now Democratic Republic of Congo (DRC) and Nigeria, more study is needed on the foreign interference effect.

The impact of governance failure on developmental outcomes is further aggravated and shaped by high levels of corruption in many resource-dependent regimes and their propensity to spend more on the military, as well as the support of resource businesses to autocratic regimes. Abundant and valuable natural resources offer opportunities for corruption and rent-seeking behaviour by the ruling elite and their cronies. This is often triggered by rapid cash inflows as a result of resource booms, with long-term economic and political consequences. Once entrenched, corruption can drastically affect the economy of a country, the quality of governance and the legitimacy of rulers. Corruption on a grand scale is facilitated by the secrecy and discretionary power of decision-makers over the awarding and development of projects, the control of economic sectors by parastatals and political cronies, or simply budgetary embezzlement. Competition between resource companies over a limited number of profitable or promising reserves raises the stakes and temptation of employing corrupt means to succeed. Even the innocuous 'wining and dining' of government officials by companies which so often characterises negotiations risks promoting a lifestyle that officials can sustain only through corruption. Windfall gains resulting from a price or production hike often trigger a 'feeding frenzy' among competing groups.[32] As was the case in several Southeast Asian countries such as Indonesia during timber booms, politicians often succeed in weakening or dismantling resource management, budgetary institutions and regulatory systems in order to capture the huge rents coming under state control.[33] Although monopolistic corruption can contribute to the maintenance of political order, competitive forms of corruption and rent-seeking among diverse agencies and levels of government often make the state in effect ungovernable. Compounded by a frequent lack of welfare-oriented fiscal policies, this generally leads to high levels of inequality and grievances undermining the legitimacy of national authorities (and their corporate associates).

The economy and politics of resource-dependent states are also affected by their propensity to spend more on defence.[34] The priority of security expenditures over social ones reflects the rulers' fears of domestic or regional opposition, as well as foreign incentives to trade resources for arms for the sake of mutually profitable political and resource flow stability. High military spending is also motivated by a 'resource variant' of the security dilemma, whereby protecting one's resource wealth motivates the increase of a defensive capacity, which is in turn perceived as a threat by potential opponents. The oil-rich Persian Gulf and, increasingly, the Caspian Sea and South China Sea are prominent examples of regions that have inspired defensive build-ups by owners of or claimants to their resources. Furthermore, the value and secrecy of arms-procurement contracts offer huge windows for corruption that often motivate disproportionate budgetary allocations. Not only is the overall economic productivity of the country affected by such military overspending, but wealth and power become increasingly dependent upon controlling rents from the resource sector and transfers to the military apparatus, raising the stakes of military control and potentially pitting military against civilian officials. Major industrialised resource-importing countries find in arms exports a convenient way to address both resource supply and balance of trade issues. This attitude has occasionally led to disastrous results, as in the case of the Persian Gulf.

Foreign diplomatic and military support of local regimes in the Persian Gulf has been closely linked to the protection of oil interests. Some of these relations have precipitated massive and violent opposition, particularly in the midst of rapid socio-cultural and economic changes. In the late 1970s, demands in Iran for greater democracy by a liberal movement merged with the aspirations of religious and traditionalist quarters for an Islamic state and led to street protests, anti-US demonstrations and hostage-taking, the exile of the Shah and, ultimately, the Islamic Revolution. American support for the Saudi regime risks a similar outcome, and may have played a part in the 11 September attacks on the US homeland.

Project profitability and contractual stability are two key bottom lines for resource businesses. This leads many companies to work with regimes with very poor developmental and human-rights records. Despite the civil strife affecting the Nigerian oil region and the execution of local activists by the Abacha regime, a Western oil company manager could bluntly argue that 'for a commercial company trying to make investments, you need a stable environment. Dictatorships can give you that.'[35] Former French parastatal oil company Elf was for decades at the heart of such relationships between French economic and strategic oil interests and several regimes in the Gulf of Guinea, with tens of millions of dollars sustaining mutually profitable partnerships between autocratic rulers, corrupt company officials and French political backers.[36] In most cases, businesses benefiting from (cosy) relationships with dictatorships will refrain from supporting democratisation processes that could jeopardise their interests – even if they prefer a democratic regime with a strong rule of law. Businesses have also been wary of 'messy democratisation processes' resulting in political instability and violence that threaten their investments. If strongly democratic regimes are more stable than autocratic regimes, the risk of instability and conflict is actually higher for formative democracies in the process of regime transition.[37] The problem, however, is that autocracies became increasingly unstable after the end of the Cold War, as a result of international and domestic pressure

for democratisation, structural adjustment and transition to market economies, as well as decreasing support from foreign sponsors. Since many
resource-dependent states were autocracies, it is not surprising that they
appeared more prone to armed conflicts.

Resources, governance failure and identity in Nigeria

Over the past 30 years, more than $350bn worth of oil has come out of the
Nigerian ground, but the percentage of Nigerians surviving on less than
$1 a day has risen from 36% to 70%.[38] Income inequality has also risen dra-
matically: whereas the top 2% earned the same total income as the bottom
17% in 1970, by 2000 it was equivalent to the bottom 55% of the population. Not only has the oil sector failed most Nigerians economically but, as
argued by geographer Michael Watts, it has also created conditions that
have 'undermined the very tenets of the modern nation-state':

> On the one hand, oil has been a centralising force that has rendered the
> state more visible and globalised, underwriting a process of secular nation-
> alism and state building. On the other, oil-led development, driven by an
> unremitting political logic of ethnic claims making, has fragmented and
> discredited the state and its forms of governance … [beyond] massive
> corruption, corporate irresponsibility or chronic resource-dependency …
> the real deception [of oil] is the terrifying and catastrophic failure of secular
> nationalism.[39]

Paradoxically, community bargaining did occur in Nigeria, but it
contributed to undermining the state through an ethnicised multiplica-
tion of local government institutions enabling the capture of rents. Since
oil started to flow into Nigerian coffers in 1965, the number of states has
increased tenfold, in part because of the allocation of a large proportion
of oil revenues through individual states. Devolution in the context of a
federal system has not significantly improved the developmental outcome
of the oil sector, but has further sharpened identity formation. Furthermore, escalating violence is a major risk in Nigeria, as suggested by the
historical example of the 1967–70 Biafra secession war (also largely motivated by local oil reserves) and the rise of heavily armed criminal and vigi-
lante groups since the end of military dictatorship in 1999. As thousands of
deaths from inter-communal conflicts in recent years would also suggest,
Nigeria's fledging democracy leaves it possibly even more vulnerable to
such divisive identity politics and calls for a drastic rethinking of the links
between oil and governance in Nigeria.

Conflicts and violence

Not only are resources shaping economic and political contexts prone to war, but resource exploitation itself may be characterised by a high degree of conflict and different forms of violence that can also participate in setting the stage or even triggering large-scale armed conflicts. Resource ownership and revenue sharing are major sources of conflicts, with many claims violently imposed upon or rejected by local communities. On the island of Bougainville in November 1989, at an environmental impact assessment meeting, mining consultants rejected, on grounds of insufficient scientific evidence, demands by the local community for compensation for chemical pollution caused by the mining operation. The focus of the assessment was the huge copper and gold Panguna mine that had provided nearly half of Papua New Guinea's export earnings for the past 20 years. Within days, using explosives stolen from the mining company, militant local landowners blew up several electric poles at the mine site and boldly demanded $11bn in environmental compensation payments, the closure of the mine and secession of Bougainville from Papua New Guinea.[40] Initially dismissed by the PNG government, their actions led to the rapid closure of the mine. Repression and a blockade of the island by the government turned the incident into a full-scale war, with possibly more than 10,000 people dying over the next decade.

Ownership and revenue allocation aside, there are many other sources of conflict resulting from resource exploitation:

- loss of local livelihoods due to land-use changes, pollution, or forced displacement;
- allocation of employment opportunities and participation in resource management;
- restriction of access to and use of resources, such as water involved in or polluted by mining activities, as well as project infrastructures like roads or electric power station;
- changes of social status, order and values within communities resulting from changes in economic opportunities and social activities, including ostentatious consumption by privileged groups exacerbating social fault-lines;
- rapid influx of population into resource regions overstretching local services and economic opportunities and leading to tensions among and between newcomers and the local population.

Conflicts may occur at all stages of the development of the project, including at the time of closure, when local settlements are under threat of becoming 'ghost towns'. The long life span of many resource-development projects means that they become parts of deeper historical grievances, often rooted in colonial practices. Conflicts over sovereign ownership of the Panguna copper mine in Bougainville started with Australian colonial legislation forbidding direct land dealings between indigenous landowners and foreigners. At the time of independence, state royalties were transferred to the government of Papua New Guinea, which returned less than 5% of proceeds to the province, sustaining a sense of expropriation among the local population.[41] Many of these conflicts go well beyond their local setting and, increasingly, involve global networks drawing in international advocacy groups such as Survival or Friends of the Earth, multinationals such as Shell, BP or RTZ, and multilateral agencies like the UN Environment Programme or the World Bank.

Not all conflicts around resource exploitation turn violent, but many do when trust between local populations, companies and local authorities is eroded. This violence takes many different forms. In the most extreme cases, national governments unleash military forces to secure resource areas, especially in politically contested areas already affected by armed conflicts. Eager to consolidate its hold on the oil region, the Sudanese army has supported in-fighting between South Sudanese militias and used aerial bombardments as well as helicopter gunship and ground attacks to kill or drive people out of oil areas that it does not fully control.[42] Transport routes as well as production areas are sensitive. Pipelines in Burma, Colombia, Indonesia, Sudan and, now, Iraq, have a history of violence. Paramilitary operations are frequently used to stop competing exploitation by illegal miners or loggers in concession areas. Prior to the conflicts that took place in the 1990s, there was a long-established pattern of violence in African diamond fields. In the 1950s, paramilitary operations took place to rid diamond concessions in Sierra Leone of illegal miners. During the 1970s, the intervention of paramilitary force of the main diamond company, MIBA, in the former Zaire led to the death of many clandestine diggers, many of whom had been authorised by customary authorities and local police. The liberalisation of the Zairian diamond sector in 1982 precipitated a vast increase in the networks of private political and military protection.[43] This in turn set the stage of further fragmentation of authority and violence in the country. To cite another example, the much-publicised conflicts over oil exploitation in the Niger Delta embraces issues such as pollution resulting from oil spills and flaring, lack of local employment opportunities and public services, forced displacement, and fire and explosion hazards, as well as brutal repression of the local population by governmental forces. Used as a justification for political and armed resistance, these issues have fed a specific form of 'petro-violence'.[44]

These issues provide increased motivation for violent conflict, superseding other expressions of grievances, such as demonstrations and land-occupation, and everyday forms of resistance such as petty theft.[45] In Chiapas, Mexico, the Zapatista movement launched a relatively peaceful rebellion in 1994 in response to an entrenched local political economy of dispossession and neglect towards indigenous communities. By staging such a rebellion, the movement sought to challenge the global neo-liberal order and to attract the attention of the government and media as a means of improving their bargaining position.[46] Violence in Chiapas was largely a symbolic means of political leverage rather than one of re-appropriation.

To sum up, in the absence of strong institutions and a diversified economy, large resource rents are likely to result in poor economic performance and governance failure that contrast against the high expectations of populations associated with a resource bonanza. Resource dependence tends to lead to a particular kind of political rule, shaping powerful but often narrow coalitions that dampen political accountability. In their quest for power, rulers often capture and redistribute resource rents at the expense of statecraft and democracy, putting their discretionary power and fluctuating rents at the core of a political order resting on clientelism. As a result, many local politicians trade economic patronage for discriminatory identity politics when resource rents collapse and autocratic rule comes under increasing pressure. Meanwhile, in their quest for profits, foreign companies often turn a blind eye to the social and human-rights records of host governments. Conflicts and violence surrounding resource exploitation projects contribute to a bleak picture among many resource-dependent countries. As resource dependence increases vulnerability to war, some resource sectors come to offer major incentives and opportunities for the pursuit of armed conflicts.

CHAPTER TWO

Resources and armed conflicts

Natural resources have gained a new strategic importance as a result of
the shifting economics of war in a post-Cold War world. Although foreign
military assistance is again on the rise as a result of the 'war on terror',
the withdrawal of Cold War foreign sponsorship in the late 1980s made
 local resources the mainstay of many war economies throughout the 1990s.
As well as financing political violence, resource revenues have served to
stoke conflict by providing a central motive to many belligerents, from the
Iraqi invasion of Kuwaiti oil fields to civil wars financed by diamonds in
West Africa.[1] Not only does resource dependence create a political and
 economic context that increases the risk of armed conflict, but whether or
not a resource is more accessible to the government or to a rebel group
may shape the likelihood and course of civil war. The practical requirements of accessing resources can influence the military strategy followed:
a coup d'état targeting the capital, or a protracted situation of 'warlordism'
established over resource-rich areas. This chapter explores the impact of
particular resources on armed conflicts. It starts by distinguishing between
different types of resources and goes on to discuss the significance of these
different resources on the type and course of conflicts.

Belligerents tend to use whatever means accessible to them to finance
or profit from war. The significance of resources in wars is thus not
systematic, but rather, often depends upon the availability of alternative options – a range of options that is often reduced by the negative
economic impact of war. Throughout nearly 30 years of military strug-

gle, the UNITA rebel movement in Angola drew on such diverse sources of support as the backing of South Africa, China and many Western-aligned countries, agricultural production schemes, international financial instruments and revenues from gold, timber, wildlife and diamonds. In the post-Cold War context, however, diamonds became its main source of revenue.[2] UNITA maintained access to diamond fields in part thanks its particular geography and mode of production, as well as a poorly regulated international rough diamond market. Alluvial diamonds spread over a vast territory proved difficult for the government to fully control, and isolated industrial diamond mines were vulnerable to large-scale guerrilla attacks. Nevertheless, despite having lost part of its control over diamond revenues, the government in Luanda was able to rely on quasi-exclusive access to offshore oil revenues. As a result, both sides benefited from a constant flow of revenue that contributed to the prolongation of a civil war that left half a million people dead and the country devastated.

The example of Angola illustrates how a resource-rich environment proved propitious to belligerents. Resource sectors are particularly susceptible to looting or extortion by rebel groups because of the fixed location of resources, often in remote areas. Many resources also require only minimal infrastructures and institutions to produce, tax or trade them. Compared to other looted goods, many resources are relatively 'anonymous' and can be laundered through legal markets. Unlike manufacturing, extractive projects cannot be relocated. Confronted by hostilities, extractive businesses will have more incentives to remain in the country than sectors like light manufacturing (although these may be sheltered from the impact of war by their urban location). Extractive businesses generally seek to sustain their operations, and protect

Resource	Accessibility by rebel forces			Price range ($/kg)
	Exploitation	**Theft**	**Extortion**	
Alluvial gems and minerals	High	High	High	20,000–500,000
Timber	Medium	Medium	High	0.1
Agricultural commodities	Medium	Medium	Medium	1.5 (coffee)
Onshore oil	Low	Medium	High	0.12
Kimberlite diamonds	Low	Medium	Medium	500,000
Deep-shaft minerals	Low	Low	Medium	2 (copper)
Offshore oil	Low	Low	Low	0.12

Table 1 – *Resource accessibility by rebel forces*[5]

their assets or commercial prospects by paying 'whoever is in power' – ranging from a few dollars to allowing a truck past a checkpoint, to multi-million dollar concessions, with 'signature bonuses' (one-off bonus payment made at the time of contract signature) or resource-collateralised loans paid in advance of exploitation to belligerents. Furthermore, resource exploitation can often be sustained during conflicts, thanks to military protection of key infrastructures and staff. Potential high returns also motivate risk-taking behaviour, especially among small-scale entrepreneurs and junior companies. Some of these may even specialise in turning high-risk situations to their advantage, by going to places where larger companies may hesitate to go because of risks to the company's reputation, or economic sanctions.

War and resource accessibility

Although a resource-rich environment may generally appear more propitious to financing and motivating rebellion, the specific characteristics of a resource, its location and its mode of exploitation can affect the balance of power between belligerents. Angola's alluvial diamonds were spread over thousands of square kilometres in the bush, facilitating rebel access. In Namibia, by contrast, alluvial diamonds can be easily found and retrieved, but they are located under vast beaches along the Skeleton Coast, an area that offers no cover to a guerrilla force. As a former South West African People's Organisation (SWAPO) fighter, now Director of Mines in Namibia, noted, 'We could not have operated there, the South Africans would have simply bombed us.'[3] The volume of revenue is also important. Over the long term, the Angolan government benefited from a larger, more secure source of revenues through the industrialised and highly concentrated oil sector than did UNITA through a diamond sector that included several industrial mines but remained dominated by the use of *garimpeiros* (freelance diggers). On average, the Angolan government netted about $2.5bn from oil, compared to an estimated $500 million in diamonds for UNITA.[4] The Angolan rebel group ultimately lost out militarily in early 2002 after losing key diamond mines to the government and facing heightened difficulties in buying and transporting weapons and fuel as it lost regional allies and UN sanctions started to bite.

Not all resources offer the same level of accessibility to revenues for rebel forces (see Table 1). In some cases, rebel groups find it relatively easy to directly control the exploitation of resources, as in the case of Angolan diamonds. Elsewhere, theft or extortion schemes are more likely means of acquisition. Rebels may pursue a zero-sum game through which resources controlled by the government are targeted for sabotage, with the aim of undermining its financial situation, increasing investment risks, or demonstrating its incapacity to maintain security. In Iraq, the oil sector is the target of both looting and sabotage. The legal character of a resource also shapes specific opportunities for belligerents. In the case of an illegal resource, a rebel group is advantaged compared to a government that risks losing its international legitimacy and associated sources of support if it engages in trafficking. In the case of a legal resource, a government has the advantage, since the market will often offer higher prices for legal resources. Financial rewards often vary between resources, with drugs being particularly attractive to rebel groups because of their illegal character and high value-to-weight ratio.

Accessibility can be assessed according to two general criteria: the concentration of resource revenues; and how easily these revenues can be controlled by the government. Revenue concentration will reflect a combination of factors, including the physical characteristics, spatial spread and mode of exploitation. Resources can be categorised as *diffuse* or *point*. The former attribute increases the difficulty of concentrating revenues, the second facilitates it.[6]

Diffuse resources are, in the main, those exploited over wide areas through a large number of small-scale operators. The high accessibility of diffuse resources makes it harder for governments to control and tax exploitation, and facilitates illegal operations. In response to the highly 'lootable' character of such resources, ruling elites have developed alternative modes of appropriation detached from the legal and institutional apparatus of the state, often through parallel (or shadow) mechanisms of control of the informal sector. Diffuse resources include alluvial gems and minerals, timber and agricultural products that are not exploited through industrial modes of production.

Alluvial diamonds, gems and minerals

Diamonds found in alluvial deposits have been eroded from kimberlite volcanic pipes. While kimberlite mines are no more than a few square kilometres large and require an industrial-scale exploitation that serves to concentrate revenues, alluvial mines are spread over vast areas and at minimum only require a shovel and a bucket for their extraction. Small, low-weight, easily concealable, anonymous and internationally tradable, diamonds found in alluvial deposits are not amenable to government control. Empirical studies suggest that regions with alluvial diamonds are more likely to be embroiled in civil war than those with kimberlite diamonds, suggesting that, at least, alluvial diamonds represent a financial opportunity that can attract rebel movements. In light of their significance in financing rebellion in Angola and Sierra Leone, diamonds have been dubbed a 'guerrilla's best friend'. Diamonds also represent a 'currency of choice' for money laundering and financing clandestine activities, because of the ease and discretion with which they can be transported and traded. Prior to the 11 September attacks, al-Qaeda operators purchased millions of dollars of diamonds in Liberia.[7] More generally, diamonds are among the favourite modes for stockpiling and transferring wealth without the risk of volatile local currencies, unreliable banking systems, or financial sanctions.

A rebel group can exert more effective control over alluvial diamond mining than governments through the use of violence. UNITA's leadership retained control in the diamond sector through a mix of financial incentives and harsh punishment (including summary executions for such offences as stealing diamonds or disrespect for officers).[8] To protect diamond-mining operations and prevent diggers from taking for themselves high-quality stones that it systematically seized, UNITA set up a special force reportedly headed by the late UNITA Vice-President Antonio Dembo. Torture and executions of diggers by UNITA soldiers or Congolese collaborators explains much about the movement's capacity to control and centralise the diamond business and to prevent its own fragmentation and corruption.

Many other valuable metals and minerals have similar 'lootability' characteristics. During the early 1990s, sapphires and rubies provided the Khmer Rouge in Cambodia and the Karen in Burma (Myanmar) with significant revenues. In Afghanistan, the late Ahmed Shah Massoud, the United Front's commander, annually netted an estimated $50m from the sale of emeralds and lapis lazuli.[9] In the former Zaire, the mining and trafficking of alluvial gold in the hilly terrain of South Kivu province sustained Laurent Kabila's rebel movement, the Party Révolutionnaire du Peuple, between the late 1960s and the creation of the Alliance des Forces Démocratiques pour la Libération du Congo (AFDL) in 1996. Gold and other alluvial minerals, such as coltan (a metal ore used in electronics, in particular, mobile phones), appeared on the balance sheet of the Congolese 'war economy' financing numerous armed groups operating in this region, from Ugandan troops to the local 'Mayi Mayi' militia.

Forests

Forest resources, mostly in the form of timber, are also among the most common resources financing rebel groups, for several reasons. First, forests are globally widespread. Second, rebel groups often base themselves in forests, which provide cover from government security forces. Third, illegal logging is widespread, as many logging companies are prepared to take risks in order to access increasingly rare and valuable old-growth timber, as demonstrated by the presence of international companies in the disputed areas of Liberia, West Papua in Indonesia and the Cabinda enclave in Angola. As in the case of diamond mining and smuggling, pre-existing illegal activities and trading networks can facilitate the operations of rebel groups. Logs, however, are not easily transported like diamonds, and often require the complicity of local or regional authorities As a Thai general commented about the conspicuous nature of logs smuggled from Khmer Rouge areas across the Cambodian border, 'We are talking about logs, not toothpicks'.[10] The exploitation of timber rests on porous borders and a high degree of collusion between rebels, governments and businesses, as demonstrated in the case of Cambodia.[11] Underpaid or opportunistic military forces deployed for counterinsurgency purposes frequently join in illicit business or allow loggers to operate in rebel-controlled areas in exchange for bribes.

Agricultural commodities

Agricultural commodities tend to qualify as diffuse resources easily accessible to rebel groups. Following the resumption of the war in DRC in 1998, coffee and cattle were among the main commodities allegedly 'drained' from areas controlled by Burundian, Rwandan and Ugandan forces and their local ally the Rally for Congolese Democracy.[12] The problem of looting aside, sustaining agricultural production during war can represent a major challenge, given the disruption of normal activity, loss of manpower and physical destruction. The relatively low price and high volume of these commodities means that they don't often play a significant role in war economies, besides being generally used as food for troops. The concentration of significant agricultural revenues can nevertheless occasionally be achieved by rebel groups, especially if pre-existing plantations are targeted and transportation infrastructure is available. In Somalia, the banana sector was integrated into the war economy, and competition for physical control of the sector, including plantation land and transport/export routes, resulted in violent skirmishes between various factions. In Liberia, National Patriotic Front of Liberia (NPFL) leader Charles Taylor received 'taxes' from major companies operating rubber plantations.[13]

Point resources are exploited in small areas by a small number of capital-intensive operators. These resources include oil and deep-shaft hard mineral exploitation, such as copper, iron and kimberlite diamonds. In many instances, such resources have a low value-to-weight ratio, which means that they must be transported in very large quantities to yield sizeable profits. These resources thus generally most benefit governments (and their allies). While this can contribute to state strength and stability, it also facilitates embezzlement on a grand scale.

Copper and oil require large-scale infrastructure and authorisation by national authorities for exploitation and international trading. These resources are thus not easily accessible to rebel groups through production. Rather, insurgents will in many cases rely on other modes of access, such as theft or extortion schemes. In Colombia, where most of the oil is inland and shipped through pipelines, the sector is believed to pay an annual total of $100m per year in protection to guerrilla and paramilitary groups, while major oil companies pay $250m to the government as a 'war tax', established in 1992.[14] To sustain extortion, guerrilla units have inflicted only limited damage, to keep the oil flowing, but as a result of more effective government military offensives, they have recently shifted to a strategy of total destruction to deny oil revenue to the government.[15]

Resources can be further distinguished in terms of how easily governments can assert legitimate and effective control over resource revenues. These two dimensions are essentially defined by the political geography of the resource. Location within a politically contested area or near a porous border, for example, can make access by rebel groups to resources easier. Two broad categories can be drawn:

- **Proximate resources** are easier for the government to control and less likely to be captured by rebels than resources in the vicinity of an area inhabited by a politically marginalised group.

- **Distant resources** are more difficult for the government to control, for example, because of their location in remote territories along porous borders, or within the territory of a political opposition group.

While the oil sector offers ease of revenue concentration, it can be difficult for a government to effectively and legitimately control. As discussed earlier, the importance of oil revenues (or their perceived importance) can exacerbate politicised identities. Onshore oil infrastructure can be vulnerable, pipelines in particular, and even offshore oil platforms are subject to protection rackets by opposition groups. In Nigeria, protests and kidnapping have been staged on oil platforms. In 1998, about 100

youths occupied a Chevron platform to protest environmental and distri-
butional issues, and to demand monetary compensation for environmental and economic grievances and jobs. A joint police and navy operation
logistically assisted by Chevron resulted in the death of two protesters.
In 1999, a small commando force of the 'Enough is Enough in the Niger
River' group kidnapped three staff and hijacked a helicopter on a Shell
platform, later releasing them for a ransom. In many instances of kidnapping, the companies had not paid in advance protection fees to the 'right
people'.[16] Ironically, the large cash transfers between some oil companies and youth groups to pacify them through 'protection contracts'
have enabled some of these groups to expand their illicit activities – for
example, by purchasing speedboats to be used in smuggling, piracy and
attacks on platforms. [17]

In short, resources are more accessible to rebel groups if they are highly
valuable, easily transported and spread over a large territory rather than a
smaller area more easily defended by government troops. Rebel access also
depends on the degree of centralisation and mechanisation of the produc-

Resource characteristics	Point	Diffuse
Proximate	**State Control/coup d'état** Algeria (gas) Congo–Brazzaville (oil) Colombia (oil) Iraq–Kuwait (oil) Yemen (oil)	**Peasant/mass rebellion** El Salvador (coffee) Guatemala (cropland) Mexico–Chiapas (cropland) Rwanda (coffee) Senegal–Mauritania (cropland)
Distant	**Secession** Angola–Cabinda (oil) Chechnya (oil) Indonesia–Aceh–East Timor–West Papua (oil, copper, gold) Morocco/Western Sahara (phosphate) Nigeria–Biafra (oil) Papua New Guinea–Bougainville (copper) Sudan (oil)	**Warlordism** Afghanistan (gems, timber) Angola (diamonds) Burma (timber) Cambodia (gems, timber) DR Congo (diamonds, gold) Liberia (timber, diamonds) Philippines (timber) Sierra Leone (diamonds)

Table 2 – *Relation between resource characteristics and types of conflicts*

tion. All these characteristics contribute to the context in which rebellion is initiated and conducted, and may also inform the strategies and tactics adopted by belligerents.

Resources and types of armed conflict

The particular characteristics of local resource sectors may influence what sort of armed conflict is pursued. Distinguished here are four main categories of armed conflict: coup d'état, warlordism, secession and rioting/ peasant rebellion – foreign intervention being an additional dimension. As with the 'resource curse', this argument is not presented as an iron law. Resource sectors will not systematically dictate the type of conflict taking place. Rather, conflict analysis should take into account the resource environment in which belligerents operate. The characteristics of local resources may make some types of conflict more likely than others, as suggested by the examples in Table 2. Some types of intervention may also be more effective than others in terms of conflict prevention and resolution, as will be discussed in Chapter 3.

Resources and coup d'états

Because point resources are generally less accessible to rebel forces than diffuse resources, with the exception of theft or extortion schemes, the best option for an armed opposition movement is to capture the state through a coup d'état in the capital city. The choice of such strategy will, of course, be informed by other factors, such as the susceptibility of the targeted government to a coup, the capacity (and international acceptability) of the rebel group, the existence (or not) of a political basis for secession that would create a new state controlling the resources. For example, since offshore oil revenues are far more accessible to governments than to rebel groups, the latter stand little chance of succeeding through a prolonged military struggle and should logically favour a coup d'état in the absence of alternative sources of finance.

The conflicts in the Republic of Congo (Brazzaville) in 1993–94 and 1997 between competing politicians – former President Sassou Nguesso, incumbent President Lissouba and Brazzaville Mayor Kolelas – demonstrate that although valid, this relationship between point resources and coups d'état is not always straightforward. The Brazzaville conflict initially represented a clear contest for state power in the context of a botched democratisation process and was exacerbated by competition for the control of an offshore oil sector representing 85% of export earnings. The fact that these conflicts took the shape of coup attempts in the capital city was in this respect predictable, and Lissouba's government should have quickly won through its control of the oil rent and associated military power. However, the 1997 war dragged on for five months before being brought to a conclusion in favour of former President Sassou Nguesso by the military intervention of the Angolan government – a long-established Nguesso ally eager to protect Angolan claims over the oil-rich enclave of Cabinda and to end the use of Congo as a platform for UNITA diamonds-for-arms deals under Lissouba.

The stalemate in Brazzaville, which destroyed a large part of the capital and left thousands dead, resulted from several factors. First, a large part of the army refused to fight for Lissouba, while others supported Nguesso, their former patron and ethnic affiliate. Second, both contenders, not just the government, benefited from access to the oil rent: Nguesso was allegedly favoured by the French oil company dominating the sector.[18] Finally, at street level, militias who supported the opposing politicians looted the capital city, leading to a rapid change in the very nature of the conflict. Urban youths on all sides used the conflict to challenge the legitimacy of a corrupt elite that had dominated and plundered the country for more than 30 years.[19] Looting became known as 'killing the pig' or 'taking a share in Nkossa [the new Elf oil field]'.[20] This form of justification echoed the devastating plunder of the Liberian capital Monrovia in 1996, when NPFL fighters hijacked their leaders' military offensive, renaming it 'Operation Pay Yourself', seeing it as a form of compensation for years of fighting 'without compensation from their leaders'.[21] Both cases illustrated how micro-level conflicts and looting can instrumentalise, or even derail, large-scale agendas.

Resources and warlordism

While rebel movements generally attempt to overthrow the incumbent regime, the existence of diffuse resources distant from the centre of power can provide an economically viable fall-back position in case of failure. This is typically the case with alluvial diamonds or forests located along border areas, hence their association with economically viable forms of warlordism whereby rebel groups create areas of *de facto* sovereignty imposed through violence. In Liberia, Charles Taylor's 1989 bid for power first targeted the capital Monrovia. Prevented from capturing the Presidential Palace by the intervention of international troops, he nevertheless succeeded in establishing his effective if not official rule over 'Greater Liberia' and took control of lucrative sectors such as timber, much of it smuggled out of the country through Côte d'Ivoire. The complicity of foreign companies and Taylor's control of key infrastructure such as the port of Buchanan even allowed for the export of such a point resource as iron ore. Taylor did not limit his resource grab to Liberia, but extended it to neighbouring Sierra Leone, where his support for Foday Sankoh's Revolutionary United Front (RUF) provided him with access to diamonds. Similarly, during the 1990s, the RUF was able, thanks to its control of diamond-mining areas as well as gold and cash crops, to sustain a guerrilla war essentially targeting Sierra Leone's civilian population. In the Philippines, the lucrative 'taxation' of logging companies has sustained many insurgent groups such as the New People's Army and the Moro National Liberation Front.[22] It is often in this context that the self-interested economic agendas of insurgents become clearest, as belligerents often settle in a mutually profitable conflict stalemate.

Resources and secessions

Most secessionist movements have a historical basis and grievances towards central authorities upon which they build their claims. Many movements, however, have emerged in the context of local resources monopolised by the central state. Arguably, opposition forces operating in the context of point and distant resources have an interest in pursuing a secessionist strategy asserting sovereignty claims over these resources. Although these resources can prove difficult, if not impossible, to access through direct exploitation, theft or extortion, their existence, or in some cases their 'mythology', is a powerful tool for secessionist political justification and mobilisation, while the prospect of future revenues provides an additional source of motivation. Diffuse resources, by contrast, can be more easily accessed by local populations who may, as a result, have fewer incentives to confront the central government.[23] The discovery of large resource reserves and the impact of exploitation can play a part in the processes of identity formation and political assertion. In Western Sahara, for example, the economic and social changes associated with the development of a major phosphate industry laid 'the basis for the rise of a modern nationalist movement, setting its sights on the creation of an independent nation-state'.[24] As Saharawis recognised the prospect of an economically viable or even prosperous country, the simplistic assumption that Morocco aimed to capture their new-found mineral wealth served to mobilise armed resistance.

Resource discoveries and exploitation can serve to renew secessionist claims, while motivating the central government to hold on the resource area with a firmer (and sometimes brutal) hand. Secessionism in the Indonesian province of Aceh is historically rooted in the existence of an independent sultanate, which prevailed until the Dutch defeated it in the late nineteenth century. Yet the formation of the Aceh Freedom Movement (GAM) coincided with the exploitation of major gas reserves in the early

1970s, and GAM's Declaration of Independence in 1976 claimed that $15bn in annual revenue was exclusively used for the benefit of 'Javanese neo-colonialists'.[25] Land expropriation and the exploitation of other resources such as timber by Javanese-dominated businesses further exacerbated the conflict. Similarly, the island of Bougainville has historical claims for separatism based on geographical and identity distinctiveness. Yet, demands for 'special status' by local politicians, including favourable funding allocations during the period of transition to independence, clearly centred on the economic significance of the island's gold and copper mine in Panguna. The secessionist conflict initiated in 1989 (detailed in the previous chapter) was related to the impact of copper mining, demands for compensation and closure of the mine, as well as a 'Government of PNG [that] is not run to safeguard our lives but rather to safeguard the few rich leaders and white men' – an analysis that resonated throughout the local Nasioi community, especially after repression by PNG forces started.[26] A Nuer fighter provided a telling explanation for the renewal of conflict after the first phase of the 1955–72 war for the self-determination of southern Sudan: 'We fought for seventeen years without even knowing of the true wealth of our lands. Now that we know the oil is there, we will fight much longer, if necessary!'[27] Before signing a peace agreement with the government in 2005 ensuring its access to the oil wealth, the Sudanese People's Liberation Army had pressured the government in the north by targeting its oil installations in the south for destruction.[28]

Resources and peasant or mass rebellion

Diffuse resources involving large numbers of producers are more likely to be associated with rioting in nearby centres of power, such as a provincial or national capital, and with support for peasant or mass rebellions involving class or ethnicity issues. At worst, neo-Malthusian arguments can be integrated into discourses of ethnic hatred to mobilise large numbers of people into committing genocide, as was the case in Rwanda in 1994. The argument here is not that genocide is more likely to be committed because an ethnically divided country relies on such resources, but that politicians can manipulate grievances experienced or imagined at the individual level within a rural sector along ethnic lines to foster mass mobilisation.

The displacement or exclusion of peasants by agribusinesses and poor labour conditions on large plantations has prompted political mobilisation and the expansion of revolutionary struggles in Latin America and Southeast Asia. In Nicaragua, landlessness as well as neglect by the state and exclusion from or marginalisation within local patron–client schemes provided fertile ground for peasant support for the Sandinista revolution in 1978–79. Yet the creation of state farms by the Sandinista regime, rather than the rapid provision of individual plots, reinforced the bonds between some landed patrons and their client peasants, rapidly increasing their support for and participation in the US-sponsored Contra movement.[29] In Côte d'Ivoire, identity politics played a major role in hostilities affecting the country since the late 1990s. In a context of democratisation and an economic downturn (precipitated by the fall of cacao prices and agricultural reforms dictated by the IMF and World Bank), the issue of citizenship (or 'Ivoirity') has literally divided the country. Although the international media focused on coup attempts in the capital, migrant workers and 'northerners' have been the targets of violence and forced displacement, and the country has been *de facto* divided in two since 2002.

Given the need for vast number of workers and the difficulty of controlling these workers over large areas, highly coercive forms of warlordism are less likely to be economically viable than participatory forms of rebellion over the long term. Conditions of slavery and control of labour can be imposed over short periods through hostage-taking. To minimise grassroots challenges, over time, the armed faction is likely to act as a 'protector' towards local populations, albeit more in the sense of a mafia than a welfare state. FARC guerrilla units in Colombia, for instance, provide protection to peasants on land holdings under its control and guarantee minimum prices for both coca and agricultural products.[30] While it has recently drifted towards criminal activities, FARC's ability to sustain peasant productiveness via a balance of threats and economic incentives has been key to the revolutionary movement's viability since its inception in the 1950s. Similarly, the expansion of the New People's Army (NPA) in the Philippines in the 1970–80s was largely based on its symbiotic relationship with a peasant population whose subsistence agriculture was threatened by agribusinesses, logging companies and mining or hydropower projects. The NPA provided an alternative to the regime of Ferdinand Marcos, which had lost all legitimacy and even presence among rural communities. Although both organisations secured most of their funds from taxation and extortion schemes related to drugs trafficking and cattle ranches (FARC), and plantations, logging and mining (NPA), they have also long benefited from some popular support.

Resources and foreign interventions

Foreign interventions occur in all types of armed conflict detailed above, and often involve indirect control over 'strategic resources' such as oil or

major mineral deposits and the protection of major commercial and strategic interests. Political scientist Michael Klare foresees a rise in foreign
interventions in the near future. Demand for raw materials is growing and
resource shortages can be expected, notably as a result of Asia's growing
mass consumerism and energy demand. Disputed oil regions such as the
Persian Gulf, the Caspian Sea and the South China Sea are of particular
concern. Even if market forces and technological progress can mitigate
some of these problems, Klare remains essentially pessimistic, given the
instability of many resource-exporting countries and the readiness of
countries claiming resources or importing them to secure their access via
military force. The approval by the US of the (short-lived) 12 April 2002
coup against democratically elected Venezuelan President Hugo Chavez
demonstrated at the very least its shallow support for democracy and
constitutional regime change in key oil-exporting countries. The US-led
invasion of Iraq has been marred by allegations of 'oil grab' motives that
have further undermined the legitimacy of this intervention, both region-
ally and internationally.

A rich resource context can also offer 'self-financing' and even lucrative
opportunities for occupying troops. The US administration sought to reas-
sure Congress that Iraq could pay for a large share of its reconstruction
expenditures. Controversially, a significant part of Iraqi assets, rather than
US funds, were used by US troops for 'winning hearts' in Iraq through
fast–track construction projects or allocated to US contractors. Successive
UN expert panels have presented evidence of the looting and business
activities of Zimbabwean, Ugandan and Rwandan troops in the DRC, esti-
mating for example that the monthly earnings of the Rwandan army were
'substantial enough to finance the war'.[31] A Commission of Inquiry set up
in Uganda under British judge David Porter cleared the government and
army of Uganda of complicity in the illegal exploitation of the DRC's natural resources, but confirmed the role of several high-ranking army officers
and recommended disciplinary action against them.[32]

External actors may also intervene in secessionist attempts by manipu-
lating local political figures into providing access to resources. In the late
nineteenth century, the discovery of gold and diamonds in the newly
created Boer republics in South Africa led both to stronger resistance to
British annexation and to a massive influx of British prospectors. The
refusal of Boer authorities to grant political rights to these British *uitland-
ers* (outlanders) led British entrepreneurs such as de Beers' founder Cecil
Rhodes to arm British settler militias, and precipitated the Boer War (1899–
1902).[33] In 1957, the French government saw its resource interests threat-

ened by the war of independence in Algeria, and organised the institutional secession of the resource-rich Sahara in the south, placing it, along with parts of Mauritania and Mali, under the direct control of Paris through the Organisation Commune des Régions Sahariennes. In response, the Front de Libération Nationale (FLN) placed the territorial integrity of the country at the top of its ceasefire negotiation agenda to ensure its control of the Saharan resources.[34] French oil interests supported the attempt of the Biafra region to secede from Nigeria in 1967; the Nigerian army only started fighting in July that year, 'more than a month after the declaration of independence but only days after Shell ... agreed to pay its royalties to Biafra rather than Nigeria'.[35] In the turmoil of Belgian Congo's independence in 1960, Belgian, British and US commercial interests, eager to secure their hold on copper mines in the province of Katanga, supported a secession led by Moise Tshombe. This led to military clashes between corporate-funded foreign mercenaries and UN troops supporting the unity of the country.[36] Diamonds, as well as other alluvial minerals such as gold and coltan, financed several rebel factions and their regional allies in the eastern DRC, resulting in a *de facto* secession in 1998–2003. This secession was accompanied by a virulent public debate over the inclusiveness of Congolese citizenship and the rights of those with Rwandan ancestry to access land and mineral resources.[37]

Influencing the course of armed conflicts
The withdrawal of foreign sponsorship at the end of the Cold War motivated most belligerents to develop alternative sources of revenue. Although other factors influenced the prolongation of wars, the availability of highly lucrative natural resources allowed some of these wars to be sustained. In resource-poor Mozambique, the cash-strapped RENAMO rebel movement relied on an intergovernmental Trust Fund established to support the peace process and adhered to the peace process in the early 1990s, whereas in Angola, UNITA and the People's Movement for the Liberation of Angola (MPLA) – flush with cash from diamonds and oil exploitation, respectively – twice returned to war. While RENAMO benefited from smuggling and protection rackets run on the flow of goods from three neighbouring landlocked countries (whose overseas trade was blocked by South Africa, which supported RENAMO), there were few resources on which to build its war economy once South Africa backed the peace process. To some extent, peace in El Salvador between the government and the Farabundo Marti National Liberation movement was partially the result of a lack of resources when US support was cut off, in contrast to Colombia, where local resources

(mostly coca) sustained the conflict between the government, paramilitary groups and insurgent movements.[38] In 1988, Khmer Rouge leader Pol Pot of Cambodia stressed the 'need to find ways to develop natural resources' to fight the Vietnamese.[39] Following the departure of Vietnamese troops in 1989, the Khmer Rouge succeeded in doing so by capturing gem mines and forested areas along the Thai border, allowing them to fight the government for six years beyond the termination of their Chinese sponsorship. Outside the framework of the Cold War, many movements have relied on resource revenues to build their military capability. In Colombia, revenues extorted by the Army of National Liberation (ELN) from oil companies through kidnappings and 'taxation' of their sub-contractors helped the movement recover from a devastating government offensive in 1973, and allowed it to grow from less than a hundred to more than 4,000 members by the mid-1990s.[40]

The accessibility of resource revenues to belligerents can prolong wars in several ways. Resource revenues accessible to the weaker party allow it to continue fighting, thereby prolonging hostilities.[41] Rebel groups may not even need to control a resource area to benefit from such revenues, but can receive 'advance payments' by businesses interested in securing future prospects. Access to resource revenues can have a significant impact on the organisation and cohesion of armed movements, and thereby on the course and duration of the conflict. The proliferation of armed factions in the Liberian conflict reflected in part 'turf wars' within some of the main factions over the control of resources. Furthermore, as a conflict drags on it can become increasingly driven by financial rewards rather than political objectives, with a growing number of belligerents motivated by economic self-interest. As natural resources become more important to belligerents, so the focus of military activities becomes centred on areas of economic significance. This can have a critical impact on the location of military deployment and intensity of confrontations (as discussed above with regard to conflict types). Complementing guerrilla strategies of high mobility, concentration of forces and location along international borders, rebel groups seek to establish permanent strongholds or areas of 'insecurity' wherever resources and transport routes are located. Government troops generally attempt to prevent this by extending counter-insurgency to these areas, occasionally displacing and 'villagising' populations (i.e., resettling them in villages along the main roads). In many cases, however, government troops join in the plunder.

The perspective of peace, and even a military victory, may not always coincide with these economic agendas. Peace agreements and peace-building efforts do not seem to succeed in a resource-rich context seem to face extra difficulties.[42] Repeated failures of peace processes in Liberia have been attributed in part to the profitability of resource looting and the immunity of rebel looters from any adverse consequences. Rebel leaders repudiated agreements to pursue their aims, feeding a cycle of fragmentation and conflict perpetuation as nominal leaders failed to rally these followers to a 'national' political project. Unsurprisingly, as in Somalia, business people previously working in the shadow of warlords often emerge as powerbrokers once the guns fall silent.[43] Having accumulated financial capital and built extensive networks, they also enjoy greater *de facto* legitimacy than most politico-military entrepreneurs and often succeed in carving out a larger area of influence, in part for not directly having 'blood on their hands', and also through the provision of jobs and other economic opportunities.

Finally, resource wealth can prolong conflict by weakening the prospects for third-party peace brokerage. The issue of access to resources can cause tensions among international players. Bilateral actors are inclined to accommodate domestic interests in order to secure commercial benefits for their corporations. In addition, the ability of the belligerents to draw on private financial flows decreases the potential leverage of multilateral agencies (e.g., the IMF the World Bank and the UN) exercised through grants and loans. In many contemporary armed conflicts, private capital inflows have assumed greater importance than foreign assistance, especially compared to conflicts in the Cold War era. Resource revenues can also subvert the motivations of external actors by motivating so-called 'private resource diplomacy' geared towards commercial, rather than peace objectives. In the case of Angola, UNITA's diamonds not only allowed the rebel movement to buy arms, but also attracted diplomatic and logistical support from regional political leaders such as Mobutu in Zaire, Lissouba in the Republic of Congo, or Eyadema in Togo.[44] On the other side of the conflict, the MPLA rapidly gained favour with major Western powers and oil companies once it was established that UNITA had lost the elections and was unable to gain power through military means. In Cambodia, the Khmer Rouge benefited from the support of Thai military and political officials who used logging revenues to finance their electoral campaigning in the mid-1990s.

While many conflicts have been prolonged by the availability of resource revenues, it does not automatically follow that wars are shorter in the absence of resources. An overwhelming concentration of revenues into the hands of one party can alter the balance of power and bring about a more rapid termination of the conflict. The Angolan government's ability to rearm and reorganise allowed it to mount a decisive military campaign between 1999 and 2002 against UNITA, which was unable to exchange its significant diamond stockpiles for arms and logistical support. Ironically, UNITA's diamond wealth may also have encouraged it to pursue a bold but ultimately self-defeating conventional warfare strategy when it was isolated by the loss of regional allies such as Mobutu and Lissouba, and more stringent sanctions imposed by the international community.[45]

Furthermore, a government's greater access to resources can motivate rebel groups to defect to the government, provide an incentive to enter peace negotiations, or lure rebel leaders to the capital to join a government of national reconciliation. The applicable caveat is that unsupervised or ill-considered wealth-sharing mechanisms can prolong wars rather than consolidate peace, even if they can be used to bring belligerents to the negotiating table. As part of the Lomé peace agreement, the leader of the RUF, Foday Sankoh, was appointed chairman of the Commission for the Management of Strategic Mineral Resources (CMRRD) and took up residence in the capital. This convenient relocation greatly facilitated his eventual arrest after RUF guards in Freetown fired at people demonstrating against the movement in May 2000, following the RUF kidnapping of UN peacekeepers.

A third way in which resources can serve to end conflicts relates to the lack of discipline and fragmentation often affecting rebel movements as a result of 'bottom up' resource flows.[46] Unless the leadership is able to monopolise the means of exchange between a resource supplier and its customers (e.g. vehicles, airports, roads, bank accounts, export authorisations, middlemen, importers), allies and subordinates have the opportunity to become autonomous through commercial or criminal activities based on local resources. The inherent risk of private appropriation can undermine trust between members of an armed group. More generally, this pattern of resource flow is likely to weaken discipline and chains of command. In contrast, when resources are fed into the conflict from outside – which tended to occur during the Cold War – leaders can maintain the coherence of their armed movements through the tight control of the flow of foreign resources to their allies and subordinates. A Khmer Rouge commander noted:

> The big problem with getting our funding from business [rather than China] was to prevent an explosion of the movement because everybody likes to do business and soldiers risked doing more business than fighting.[47]

To prevent such 'explosion', or fragmentation, the Khmer Rouge fully supported soldiers and their families, and tightly controlled cross-border movements. The regime hierarchy also supervised business dealings by local units. In 1996, a new commander was sent to sort out a local Khmer Rouge unit that had become too cosy with provincial authorities as well as to settle a contractual dispute between a commercial logger and the movement; 14 workers and the boss of the logging company involved were captured and beheaded. Tensions arose because of financial disparities between Khmer Rouge regional units, aggravated by the decline of gemmining revenue caused by reserve depletion and government attacks on mining sites. This led to the movement's fragmentation, massive defections and its ultimate demise. Additionally, logging revenue dried up as the Thai government, under pressure from a British NGO, Global Witness, and the US administration, closed its border to trade with the Khmer Rouge. The fall of logging revenues from mid-1995 onwards increased tensions and distrust within the leadership. In the south western part of the country – a Khmer Rouge-controlled area endowed with gems and rich in timber – local commanders resisted demands by the party elite in the north for more hardline policies and increased revenue transfer. Both sides of the divided movement accused each other of embezzling money; the crisis led to infighting and atomised negotiations with the government. Local Khmer Rouge commanders who defected to the government obtained partial control of their territory and its resources, as well as tax exemptions on fuel imports and financial aid to build schools. Government cronies, in return, set up several casinos in the semi-autonomous territories in partnership with former Khmer Rouge along the Thai border.

Finally, an armed group that exploits natural resources is vulnerable to losing popular support and political legitimacy if its adversary portrays the group as mere bandits or criminals driven more by economic self-interest than by political ideals. When international security institutions such as the UN Security Council ignore similar 'criminal' practices on the part of government officials or paramilitary groups, this facilitates the sanctioning and political isolation of rebel movements like the RUF, UNITA and the FARC. Such a policy can, of course, run the risk of marginalising a political resolution of the conflict in favour of a military solution. This, in turn, risks alienating local populations who would bear the brunt of a conflict and drastically reducing their political support for a government appearing to prolong the war.

The influence of resources on the level of hostilities and their impact on population is ambiguous. Although more revenues should mean an aggravation of hostilities, resulting in increased impact on local populations, this is not always the case. On one hand, resources can intensify confrontations, especially around areas of economic significance. On the other, armed groups can settle for a 'comfortable stalemate' in which opposing parties can secure mutually beneficial deals to produce and market resources. This relationship can favour localised peace agreements and defections, when local commanders lower the intensity of conflict and even negotiate their individual disengagement without approval from their supposed leaders. In Cambodia, the Khmer Rouge benefited from export authorisations granted by the co-Prime Ministers Norodom Rana-riddh and Hun Sen to Thai companies operating in its areas. The Khmer Rouge also granted logging companies permission to operate chainsaws, lorries and sawmills in areas that it controlled in exchange for monthly fees in cash or rations. These fees were crucial to the viability of small Khmer Rouge groups far from their bases.[48] Government soldiers and authorities were able to collect fees from the same loggers. In the short- and medium-term, neither side had an incentive to change the status quo.

Large revenues can allow belligerents to shift from a war of terror target-ing 'easy targets' such as civilians to a conventional type of conflict, possibly more respectful of the laws of war. In contrast, desperate belligerents lacking access to resources may well intensify attacks on civilian populations. There seems, however, to be no clear correspondence between access to resource wealth by belligerents and attacks on civilians. Rebel groups benefiting from large resource revenues, such as RUF and UNITA, have committed atrocities against civilians. Furthermore, many civilians often use the context of war to engage in violent and economically motivated crimes.

In sum, natural resources can influence the course of conflicts by moti-vating and financing belligerents. Although there is no deterministic rela-tionship, resources can serve to shape the type of armed conflict taking place, the territorial control objectives, the duration and intensity of the conflict, and relations between belligerents and populations. Resources can also affect the internal cohesion of armed movements and motivate collusion between adversaries, especially when exploitation or trading requires such partnership. As such, the context set by resource depend-ence and the influence that resource sectors may have on the course of armed conflicts have significant implications for both conflict prevention and conflict resolution.

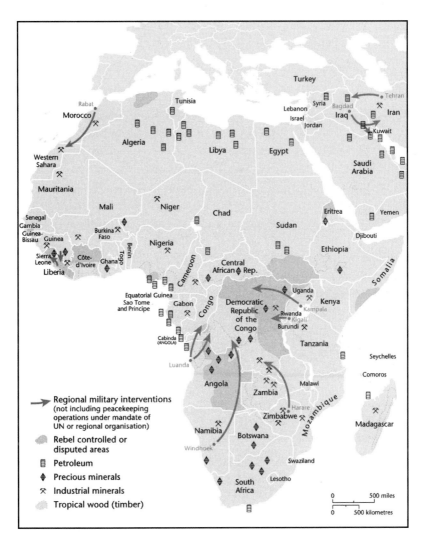

Map 2 – *Natural resources and armed conflict in Africa and the Middle East*

Chapter Three

Implications for conflict prevention and termination

Freed from Cold War confrontation and shaped by new norms of rights and sovereignty, the current international environment is more conducive to the promotion of better resource governance, which will contribute to peace. So far, efforts have been mostly concentrated on the protection of the environment and upholding the human rights of indigenous communities directly affected by resource exploitation. Preventing and ending armed conflicts related to resources, however, requires broader engagement by external actors, more transparent and accountable management of resource sectors, and the imposition of specific measures for tackling the role of resource revenues in fuelling war.

Addressing the resource curse

A broad policy framework is required to ensure that resource wealth is captured and diffused in the interest of local populations and security. This framework requires the participation of coalitions of powerful and diverse interests to pursue several objectives.

- maximising and cushioning resource revenues;
- allocating revenues fairly and efficiently;
- diversifying resource-dependent economies; and
- promoting peaceful and secure resource supply.

Maximising and cushioning resource revenues

Resource-exporting countries should promote higher revenues and less volatile prices for raw materials. Revenues should be increased not by increasing the volume of production, which tends to depress prices, but through a combination of improved productivity, demand growth, improved taxation of resource sectors and the adding of value to exported commodities. Governments and businesses in producing countries often accumulate reserves during boom times to be used as revenue-stabilising funds, released when the economy or commodity prices flag. Such funds can also ensure greater equity between regions by allowing for national revenue redistribution, thereby acting as a mechanism of national solidarity and equality. Savings funds can also 'sterilise' rent windfalls, preventing uncontrolled and wasteful public spending. Such funds, however, are prone to being raided by politicians to serve corrupt or populist motives. A simplistic strategy of privatisation of resource parastatals is not an easy way out either, as demonstrated by the disastrous 'Zairianisation' of the Congolese economy, when it was nationalised under Mobutu in 1973, or the corrupt insider privatisation process in Russia in the early 1990s. In both cases, the politicisation of privatisation and inadequate institutional frameworks for the newly privatised entities led to asset stripping and capital flight, with disastrous economic and social consequences. In Côte d'Ivoire, the government-controlled coffee and cacao stabilisation fund (CAISTAB) allowed for producers throughout the country to obtain similar prices, whatever the distance from the ports in the south. This measure increased farming revenues in the otherwise impoverished northern part of the country. The CAISTAB contributed to the consolidation of the 33-year rule of late President Houphouet-Boigny, but it also delayed economic adjustments that aggravated the economic situation of the country. The privatisation and subsequent liquidation of the CAISTAB in 1998 jeopardised the Ivoiran economy, sent the cacao price tumbling, and aggravated tensions between an affluent urbanised and Christian south, a mostly Muslim north, and a migrant labour force accounting for close to a third of the population.

At an international level, resource prices need to be stabilised to favour poor exporting countries rather than rich importing countries, who often maintain effective trade barriers and subsidies to protect their economies. Most international commodity agreements and organisations have collapsed over the last 15 years, in part because superpower use of these schemes during the Cold War to sustain favourable regimes in producing countries has been superseded by the predominance of neo-liberal economics. The main reason was a general unwillingness to resolve the practical difficulties of price stabilisation in a world perceived as increasingly market driven, globalised and competitive.[1] In 2000, the IMF eliminated its Buffer Stock Financing Facility, noting that there was no eligible commodity agreement in which this credit scheme could be used. The IMF also reduced its provision of non-concessionary medium-term loans for covering temporary shortfalls in export revenues through the Compensatory and Contingency Financing Facility. Similarly, while the European Union (EU) remains interested in boosting these stabilisation mechanisms, its STABEX scheme was abandoned because it entailed excessive delays between initial economic shocks and the disbursement of project-specific financial aid.[2] Rather than following its own market-oriented policies, the World Bank currently only promotes financial-hedging techniques through a couple of pilot projects among a few commodity producers.[3]

As a result, resource-dependent countries tend to be 'price takers' rather than 'price makers', and are fully exposed to volatile and generally decreasing prices. Because resource price shocks cannot be covered domestically, given their magnitude and the absence of alternative sources of growth, exporting countries require protection against the collapse of export prices. When helping to define the post-war economic order, John Maynard Keynes proposed the creation of an international institution in charge of stabilising commodity prices, but the US Congress opposed the idea. The International Trade Organisation was stillborn and its new embodiment, the World Trade Organisation, has so far largely failed to deal with this issue.

Fair pricing mechanisms and compensatory financial instruments are needed at both international and domestic levels. In the short term, emergency measures are required in sectors where prices have already collapsed. About 25 million workers worldwide, the vast majority of whom are poor peasants, rely on the coffee industry for their livelihood. Members of the International Coffee Organisation should, for example, impose an export fee that could be used to alleviate poverty among coffee workers. Maximising and cushioning resource revenues for exporting countries, however, is not a panacea. Greater resource revenues may further reinforce resource dependence, and it is therefore crucial that fair revenue allocations, as well as sound institutions and economic reforms, consolidate such gains.

Allocating revenues fairly and efficiently
The greatest challenge in regulating the use of natural resources is to guarantee that revenues benefit populations and strengthen rather than erode the quality of governance. Transparent, negotiated and accountable revenue

collection and allocation schemes are essential. Regularly audited funds channelling resource rents are increasingly used, but the details of such audits are rarely made public and transparency is rarely sufficient, given the myriad opportunities for corruption, embezzlement and mismanagement available to government officials and businesses. Even when audits and figures are publicly available, potential auditors from civil society or elected representatives often lack the capacity, time or funds necessary to investigate revenue management properly. In many developing countries, the IMF is the only external public auditor, but it often follows its own agenda of macro-economic and monetary stabilisation, is adverse to engaging in political issues and follows confidentiality rules that may improve its own data access but undermine accountability by limiting public disclosure.

Effective schemes for keeping track of resource revenues require constitutional and legal reviews, as well as the creation or strengthening of institutions that can provide checks and balances on government officials and businesses, such as anti-corruption regulations and organisations, public auditing and independent judiciaries. Such institutions need to be effective at multiple levels, from initial contracts on resource projects to the disbursement of public budgets by local authorities. Decentralisation of revenue management can favour greater accountability, but involves a shift in power from central to local authorities, with potential risks that could include a loss of national cohesion, heightened local corruption and less effective management. In this vein, the political and economic outcomes of the Indonesian government's recent fiscal devolution of 70% of energy and mining revenues to the provinces of Papua and Aceh are likely to be highly relevant.

To curtail corruption, the Organisation for Economic Cooperation and Development (OECD) and the EU have adopted anti-corruption regulations to be imposed on both aid recipients and donor countries' corporations. The EU introduced into the Cotonou agreement a clause of 'essential elements' for partnership, such as respect for human rights, democratic principles and the rule of law, and the absence of 'serious cases of corruption'.[4] Following the suspension of EU aid to Liberia in June 2001, the government of Liberia submitted to the EU's demand for an audit of Liberia's public finances. The IMF and the World Bank are also taking bolder steps to tackle budgetary transparency and corruption issues. New NGO networks with an international reach, such as Transparency International and Global Witness, are now focusing on corruption issues. The active collaboration of the finance and resource industries is essential: their payments to state or other recipients need to be transparent; and they can exert vital leverage over the management of public revenues they help create. Rather than accepting clauses of confidentiality over taxation issues in their contracts with governments, the international norm should be for businesses to push for minimal standards of transparency and account-ability. Such norms could be enforced by stock exchange regulations and standards of corporate governance. While capital-market sanctions remain controversial, improved norms of mandatory corporate disclosure have been high on the Comprehensive Corporate Reform Agenda of the US administration, formulated following the accounting scandals of Enron and other major publicly listed companies in 2001. The pledge by US President George W. Bush in 2002 to 'move corporate accounting out of the shadows'[5] should not only benefit shareholders in the US, however, but also populations in resource-producing countries.

Debt relief should figure high on the question of fair and efficient reve-nue allocation. Many resource-exporting countries use a large part of their revenues to service their debts, exposing repayment to volatile sources of income and detracting from more productive uses. Debt relief, governance and commodity prices should be linked to help resource-dependent coun-tries move out of the resource trap. Low resource prices should automati-cally reduce debt repayments and suspend accruing interest. Debt relief should reward improvements in governance and economic diversification. The principle of 'odious debt' (according to which a debt contracted under a repressive dictatorial regime should not obligate future servicing) should be applied to cancel debts following successful political transition. This mechanism could be funded partly by resource companies that have bene-fited from commercial partnerships with past brutal dictatorships. Such a policy would reduce incentives for foreign companies to work with similar regimes in the future.

A drastic form of resource revenue allocation that prioritises the needs of local populations would be to distribute revenues directly to the people. Indirect disbursement through the payment of dividends derived from an oil-savings fund is already implemented in Alaska (the 2004 amount was $919.84 per resident, including children). Such savings funds, however, are often subject to being raided by politicians, so direct disbursement may be preferable. While such direct disbursement would have a positive impact on individual incomes, it would negatively affect the fiscal position of the state. In this respect, governance failure resulting from a lack of taxation – and its impact on accountable representation – could be addressed by partly taxing the disbursed revenue as part of income taxation. Many critics of such an approach, especially among the international aid sector, fear that cash provi-sion is likely to fail in most developing countries, but some experiences have been positive, notably in Mozambique. Although such a scheme would face many challenges, it has been vigorously advocated by some human rights activists and academics in the case of Iraq and Nigeria, in particular.[6]

Diversifying resource-dependent economies

Resource-dependent countries need to diversify their economies to provide a broader range of income. Such an effort can take the form of vertical diversification through the processing of resources into valued-added products, or horizontal diversification into non-related sectors. Achieving economic diversification is a major challenge, and numerous attempts by resource-dependent countries have failed or even proved counterproductive. Domestic challenges include a dearth of financial markets, low productive and marketing skills, and political impediments to licensing and taxation. At the international level, competitiveness in manufacturing has long been obstructed by importing countries imposing higher tariffs on processed than on raw materials. Many OECD countries have taken specific steps to open their markets to non-WTO and poor resource-dependent countries. In 2001, the EU initiative 'Everything But Arms' opened up the European market to nearly every product from all Least Developed Countries (LDCs), with the exception of Burma. The hope is that greater access for secondary commodities will diversify LDC economies. However, the track record of such preferential agreements, already in place for the African, Caribbean and Pacific group of countries (ACP) since the 1960s, is, to say the least, disappointing. Of all ACP countries only Mauritius – which has very few natural resources – represents a clear success.

Promoting peaceful and secure resource supply

Promoting secure resource supply for importers can reduce tensions and foreign interventions. As discussed earlier, the challenge is to balance the interest of importers with that of local populations and, in some cases, find alternatives to the deployment of military force or complicity with local dictatorships. There is also a need for international institutions to manage the demands for key commodities. Petroleum is an obvious candidate in this regard, having already been used as an economic weapon and proved a major source of instability, and having motivated military action by major powers in the past. Supply diversification and technological substitutes are also part of the solution, but they can hurt producing countries by decreasing demand and prices for resources – and thereby revenues – and call for negotiated solutions to reconciling the needs of exporters and those of importers.

Resource interdependence can help foster stability and the transfer of wealth towards poor resource exporters. An oft-cited example is the interdependence that results from the development of a gas sector dependent on fixed pipelines: countries connected in this way are motivated to collaborate and put aside their political differences. Unlike oil or other resources, gas cannot easily be shipped elsewhere, nor can importing countries find as cheap an energy resource. The case of the Burma–Thailand gas pipeline, however, proves that economic interdependence can also sustain dictatorial regimes and fail to address contentious political issues, such as border control and drug smuggling, which result in military skirmishes.

Importing countries have a responsibility to ensure that resources are exploited legally. Not only do illegal activities have fiscal and often environmental consequences in exporting countries, but illegal trading channels are frequently used to launder 'conflict resources'. About a third of the timber imported by the G8 and China, for example, has been illegally felled.[7] Rather than setting an embargo on timber coming from countries where illegal exploitation is prevalent, the EU, for example, is trying to include legality requirements in procurement rules and is pursuing reform in the forest industry to improve the situation. The example of the German-based Forest Stewardship Council scheme, which certifies timber has been sustainably harvested, is relevant in these matters. National customs authorities and the World Customs Organisation can help to track conflict resources, in particular, through its network of Regional Intelligence Liaison Offices. Specialised industry associations, the UN Conference on Trade and Development (UNTACD) and businesses involved in trans-border trade verification and certification can also assist in this task.

Ending 'resource wars'

The scope and number of measures to curtail the access of belligerents to financial means has greatly increased since the end of the Cold War. The following section examines the broad options available, and then more specific measures such as targeted sanctions, international investigations by the UN and civil-society organisations, the use of judicial trials and aid conditionality, market regulation and commodity certification schemes, as well as economic supervision.

Capture, share, or sanction?

Three broad types of initiatives have been taken to end resource wars: capturing resource areas from rebel forces, sharing revenues between belligerents and imposing economic sanctions. If none of these initiatives provides a comprehensive solution on its own, their relative effectiveness appears to vary according to the characteristics of the resources that are being targeted.[8] Diffuse resources (most accessible to rebel forces) are best addressed through sanctions, while military capture is most effective in the case of point resources. Controversially, when illegal resources are financing war, like narcotics in Burma, sharing arrangements between belligerents prove to be more effective; but it is arguably a rare official option for governments and even less so for external actors attempting to resolve the conflict.

The military capture of resource areas from rebel forces appears to be a deceptive quick fix: successfully implemented, it often forces the targeted party into a settlement, but ultimately fails to usher in a stable peace. Military capture requires significant follow-up to avoid the recurrence of hostilities. Sharing resource revenue between belligerents can be as successfully achieved as military conquest, and is more rapidly followed by conflict settlement, but rarely results in a stable peace. This finding, however, may reflect a timing issue, since agreement on revenue-sharing is often part of a conflict settlement. Given the asymmetry between belligerents and the risks of duplicity underlying many of these sharing agreements, third parties may have a role as guarantors (see below). Adequately mandated peacekeeping forces and an international supervising mechanism for the resource sector can help provide such guarantees. Sanctions have a poor overall record in terms of implementation over the past 15 years (1989–2004), but major improvements have been noted since the late 1990s in terms of monitoring and enforcement. Furthermore, sanctions are generally lifted only once a conflict is comprehensively settled, so it can be said that they do have a role in contributing to a lasting peace.

Policing and shaming: UN sanctions and expert panels

Under Article 41 of the UN Charter, the Security Council may impose restrictions on economic relations by UN members with targeted countries or groups. While sanctions have generally being used as an economic leverage to promote negotiations or policy change, UN sanctions regimes have increasingly aimed to put targeted belligerents 'out of business' by prohibiting commodity exports upon which they rely economically. The logic of sanctions has thus evolved from containment and influence to policing. Prior to 1990, only Southern Rhodesia had been subject to a commodity export embargo. Since then, seven countries or armed groups have had their commodity exports prohibited or otherwise limited (see Table 3).

To enforce sanctions, the Security Council relies on its members to prohibit trading in target commodities through regulatory or military

Year	Country	Resolution
1966	Southern Rhodesia	S/RES/232 (1966) and 253 (1968) on all commodities.
1990	Iraq	S/RES/661 (1990) on all commodities; S/RES/665 (1990) calls for halt, inspection and verification of all maritime shipping in the Gulf area to ensure strict implementation of S/RES/661.
1991	Yugoslavia	S/RES/757 (1991) and 787 (1992) on all commodities.
1992	Cambodia	S/RES/792 (1992) on log exports, requests adoption of embargo on minerals and gems exports, and requests implementation measures by UNTAC.
1993	Libya	S/RES/883 (1993) banned the provision to Libya of equipment for oil refining and transportation.
1994	Haiti	S/RES/917 (1994) on all commodities.
1998	Angola	S/RES/1173 (1998) on all diamonds outside governmental Certificate of Origin regime and the provision of mining equipment and services to non-government-controlled areas; S/RES/1237 (1999) establishment of expert panels; S/RES/1295 (2000) establishment of a sanctions-monitoring mechanism.
2000	Afghanistan	S/RES/1333 (2000) banned the provision to Taliban-controlled areas of acetic anhydride used in heroin production.
2000	Sierra Leone	S/RES/1306 (2000) on all rough diamonds pending an effective governmental Certificate of Origin regime, and creation of expert panel on the implementation of sanctions.
2000	DR Congo	S/PRST/2000/20 establishment of expert panel on the illegal exploitation of natural resources and other forms of wealth.
2001	Liberia	S/RES/1343 (2001) on all rough diamonds, and establishment of an expert panel; S/RES/1408 (2002) establishment by government of Liberia of transparent and internationally verifiable audit regimes on use of timber industry revenues; S/RES/1478 (2003) on all timber exports.

Table 3 – *UN Security Council sanctions against resource exports*

means. According to Article 25 of the charter, members must abide by and implement these sanctions. Although member states may themselves be sanctioned for non-compliance, there is general reluctance to propagate so-called 'secondary sanctions'. Only Liberia, under Charles Taylor, came under such sanctions for its role in supporting the RUF rebellion in neighbouring Sierra Leone and threatening Guinea. Furthermore, the lack of adequate national legislation and enforcement by some governments has left many sanctions with a merely rhetorical effect, such as that against the

Khmer Rouge in Cambodia in the early 1990s. Such ineffective enforcement can favour criminal groups by forcing more legitimate companies out of resource sectors, with potentially negative consequences in terms of conflict prevention. Recognising this dynamic in the context of sanctions against UNITA, the Security Council specifically urged 'all States … to enforce, strengthen or enact legislation making it a criminal offence under domestic law for their nationals or other individuals operating on their territory to violate the measures imposed by the Council'.[9]

Very few implementation measures have involved ground troops. Immediately after the Iraqi invasion of Kuwait in 1990, the Security Council imposed a sanction regime to block Iraqi-controlled oil exports. The Multinational Interception Force (MIF), led by the US Navy, acted under Security Council Resolution 665 (1990) to interdict all maritime traffic to and from Iraq to ensure a strict implementation of sanctions. When these sanctions were imposed, Iraq was highly dependent on foreign trade and had huge debts. As a result of the sanctions, Iraqi oil exports were reduced by 90% between 1990 and 1995, crippling its economy. Yet Saddam Hussein proved sufficiently resilient or unconcerned by the plight of his population to withstand the sanctions, which were progressively eroded by the Oil-for-Food Programme and oil smuggling; for example, Iranian authorities sold 'transit permits' to smugglers to pass through their territorial waters.[10] Furthermore, political 'realism' led the US to avoid a confrontation over smuggled oil with 'friendly regime' such as Jordan and Turkey which were importing vast quantities of smuggled Iraqi oil (the US itself was one of the main importers of officially exported oil). By the late 1990s, smuggling was estimated to bring annual revenues in excess of $500m. Major as well as small oil purchasers and companies were involved. Nevertheless, there remains some potential for more effective enforcement of sanctions. The unique physical characteristics of oil fields and the availability of databases allow for the identification of the oil transported to the international market. Shell, for example, was fined $2m by the UN for such dealings after one of its tankers was found transporting Iraqi oil in April 2000.[11]

In 1992, the Security Council asked the UN Transitional Authority in Cambodia (UNTAC) to impose a log export ban to dampen the Khmer Rouge's efforts to undermine the peace process. The UNTAC force was 22,000-strong but although extensive, its mandate did not include the use of force. Its leadership was wary that any military move to control the logging sector could get out of hand, given the extensive profits at play. Instead of confronting illegal logging and possibly requesting authorisation for the use of force from the Security Council, UNTAC deployed military observers without power of arrest to monitor Cambodia's international border. The Khmer Rouge refused to allow monitoring in its territories, and the Thai government denied access on its side of the border. Despite holding documentary evidence of the log trade into Thailand from Khmer Rouge territories, UN Secretary-General Boutros Boutros-Ghali took a 'quiet diplomacy' approach towards the new Thai government, which had just replaced a military junta. Even though he took the matter personally to the foreign minister of Thailand, the Thai government remained reluctant to act against pro-Khmer Rouge Thai military and provincial officials who profited from ties with the rebel group. Although all log exports were eventually prohibited by Thailand, the Cambodian government's lifting of the ban after nine months undermined its overall impact. In the end, Thai timber imports for 1993 were only 20% lower than in 1992 and 1994, while the gems trade, which also benefited the Khmer Rouge, continued largely unabated. As in Iraq, the ban consolidated corruption and the control of illicit economic activities by political actors.

Besides physically denying the sale of commodities through military means, Security Council compliance measures have mostly targeted the identification and criminal conviction of sanctions-busters. Following the past UN failures in Angola, including its failure to implement sanctions, lobbying by NGOs and the urging of Canadian officials in 1999 resulted in new approach under which international agencies, governments and businesses were directly enlisted to help implement sanctions. Along with media reporting and advocacy work by NGOs such as Human Rights Watch, Global Witness and Partnership Africa Canada, the UN publicly 'named and shamed' sanctions-busters, including heads of states, shattering the law of silence generally characterising relations within the UN. Key to this revolution was the innovative involvement of independent experts whose work was not hampered by diplomatic protocol and who were given both the time and budget to conduct in-depth investigations. The main enforcement strategy of sanction became public 'naming and shaming'. As put by the chairman of the UN expert panel for the DRC, the goal of this strategy was 'to shock' countries the panel suspected of benefiting from the 'looting' of eastern DRC.[12] The presidents of Togo and Burkina Faso were among those 'named and shamed' by the panel on UNITA sanctions. The first reaction from Rwanda and Uganda was to protest and even threaten disengagement from the Lusaka peace agreement; however, a more conciliatory tone was adopted in the medium term, and may have contributed to the subsequent realignment of Ugandan-backed rebels with Kinshasa. Critics of this policy have argued, however, that 'naming people without shame' is useless in light of continued predatory practices by elite networks in the region.[13]

UN expert panels do not work under cover and rely on voluntary testimonies, which complicate their task in comparison to intelligence or police work. Greater cooperation from intelligence agencies and the financial sector is needed in this regard. Even if their most sweeping recommendations have not been implemented, UN expert panels have been able to provide information that has significantly affected the operations of sanctions-busters, by indirectly curtailing access to financial credit, forcing them to change their logistical base or to seek protection in friendly countries. With staff hired on a consultancy basis from a pool of independent experts, and budgets averaging $1m for a typical team of five working over a six-month period, expert panels are relatively cheap and easy to set up. Adding Interpol representatives facilitates exchanges with police institutions worldwide. There is some momentum on the part of the UN Secretariat, member states and policy think-tanks behind the prospective creation of a permanent sanction-monitoring unit that would centralise information and logistically facilitate panel work and links with relevant institutions. The institutional location of such a unit under the Security Council or the Secretariat remains debated, however. Placing it under the umbrella of the Security Council could prevent any undermining of the credentials of the Secretary-General's 'good offices' by expert panels denouncing high-level officials, but the legitimacy of the Security Council is itself somewhat tainted by the dominance of the five permanent council members. Another possible model is that of parallel independent monitoring of the humanitarian impact of sanctions, as occurred in Liberia. Poor reporting and broad criticisms against the role of specific Western companies have occasionally eroded the credibility of, and support for, the expert panel mechanism among UN Security Council members. Expert-based and independent investigation into war economies and the implementation of sanctions should become a permanent feature of the international security apparatus.

Individual countries and regional organisations have also imposed commodity export sanctions. In the US, the federal administration, state governments and even municipalities have multiplied unilateral sanctions against countries or individuals – sometimes with extra-territorial reach, as through the Iran–Libya Sanctions Act (1996). The US administration was among the first to impose sanctions against Iraq immediately after its invasion of Kuwait. It targeted the military regime in Burma in 1995, through an investment moratorium affecting mostly US energy companies. But that measure fell short of requiring disinvestments or even deterring reinvestment in existing projects. In the context of civil strife and the hanging of Ogoni activists by the Abacha regime in 1995, a 'Nigerian Democracy Act' banning new investments was presented to Congress, but was rejected after lobbying from US business associations and the Nigerian government. President Bill Clinton offered to pass equivalent legislation if multilateral consensus could be achieved, but European governments opposed the move on the ground that it could have jeopardised debt repayment and risked expropriation of oil-business assets, and preferred to adopt non-economic sanctions.

Commodity exports have also been the focus of sanctions by regional organisations. The Economic Community of West African States (ECOWAS) imposed economic sanctions against NPFL-controlled areas in Liberia in 1993, after its military arm, the ECOWAS Monitoring Group (ECOMOG) had organised a military blockade and takeover of Taylor's leading port in Buchanan, from which the NPFL had imported arms and exported timber, rubber and iron ore. Although Taylor lost a significant portion of his income as a result, unimpeded diamond and timber trafficking allowed him to maintain his military strength until he was elected president in 1997. A political and trade embargo was imposed for more than two years on Burundi by neighbouring countries after a military coup in 1996. Much criticised for its macro-economic and humanitarian impact, the embargo was systematically violated by participating states, allowing for the export of all key commodities from Burundi, notably the smuggling of coffee, which was also encouraged by the low prices offered by the parastatal marketing board.[14]

Sanctions have rarely proved effective in the short term. Furthermore, they often have a negative humanitarian impact on local populations. Yet the idea that commodity export sanctions should be avoided or lifted to promote positive economic engagement that benefits populations does not easily hold with respect to extractive industries. This is primarily because the targeted belligerents often control the rent of the extractive sector, and secondarily, because extractive industries generate little local employment. In Angola, for example, the oil rent is controlled through the presidency, Ministry of Finance, and the parastatal oil company Sonangol. The sector has generated only about 10,000 local jobs and very little revenue 'trickles down' to the general population. UN sanctions against the Liberian logging sector under UNSC Resolution 1478 have been heavily debated: UN expert panel members and Global Witness claim that timber revenues supported the RUF, while the humanitarian coordinating branch of the UN (OCHA) argues that too many jobs would be lost in an already very weak economy.

France and China – the two main importers of Liberian timber – initially opposed sanctions, taking their cue from the humanitarian argument and demanding undisputable evidence of the link between logging and arms, but eventually agreed to sanctions.

Restricting market access and ensuring financial transparency

Financial markets and individual businesses involved in the processing and marketing of conflict resources are under growing pressure to end their complicity in war economies. Sanctions, in this respect, are not the only tool with which to restrict belligerent access to resource markets. A disinvestment campaign by a coalition of labour unions and human-rights activists successfully lobbied US fund managers and investors about the involvement of PetroChina in Sudan and Tibet, reducing the capital raised by the company through its New York Stock Exchange in 2000 by an estimated US$2 billion. In 2001, the US House of Representative passed the Sudan Peace Bill, which calls for the president to prohibit 'any entity engaged in the development of oil or gas in Sudan from raising capital ... or from trading its securities ... in any capital market in the United States'.[15] The final Act signed by President Bush a year later does not include such precise prohibition, but refers to 'all necessary and appropriate steps to deny the Government of Sudan access to oil revenues to ensure that [it] neither directly nor indirectly utilizes [them] to purchase or acquire military equipment or to finance any military activities'. The Act nevertheless sent a message to oil companies in Sudan already affected by disinvestments from US pension funds such as Talisman, which sold its Sudanese operations to an Indian consortium shortly after the Act was signed. By discouraging the public listing of state or privately owned companies, however, such campaigns could prevent the adoption of higher standards of accounting and disclosure. Listing on the London Stock Exchange, for example, requires companies 'to create systems to identify, evaluate and manage their risks and to make a statement on risk management in their annual report'[16] that account for business probity issues. The UK Pensions Act requires pension funds to inform customers about social policy commitments and to lay out ethical considerations in their investment portfolio. Most financial markets, however, continue to provide funds without stringent regulations on corruption and the financing of war.

The 'Publish What You Pay' campaign organised by a coalition of NGOs led by George Soros and Global Witness aims to make the disclosure of all payments to host governments by oil, gas and mining companies a mandatory condition for listing on international stock exchanges. Such regulatory measures should eliminate concerns by companies about confidentiality clauses imposed by host governments and level the playing field between competing listed companies. Although such a measure would secure the compliance of most major resource companies, it would not affect non-listed companies, such as privately owned or state companies. This loophole could be avoided by similar obligations of disclosure under national or regional company law. A further potential loophole concerns the regulation of international brokers registered in offshore jurisdictions and parastatal companies in host countries. Such brokers specialise in the dirty work of getting resource concessions through corrupt deals, before selling them on in a 'clean' manner to larger resource companies who turn a blind eye to their provenance. Local 'sleeping partners' associated with the operations of resource companies, in the form of board directors or parastatal companies in charge of some subcontracting operations, also act as vehicles for corruption by scooping large cash bonuses, commissions, or profit shares. The 'Publish What You Pay' campaign caught the interest of British Prime Minister Tony Blair, and the British government launched its own Extractive Industries Transparency Initiative (EITI) in 2002 seeking to 'ensure that the revenues from extractive industries contribute to sustainable development and poverty reduction'.[17] The EITI lays out standards of transparency, accountability and prudent management of resources principles and provides reporting guidelines for voluntarily participating countries and companies.

Individual companies have also come under pressure to disengage from war-torn areas and pariah states through NGO campaigning and non-regulatory governmental pressure. Sabena/Swissair suspended their shipment of coltan by cargo flights from eastern Africa after being named by a UN expert panel report as intermediaries in the coltan trade fuelling the war in eastern Congo. This decision was also influenced by an innovative campaign by Belgian NGOs denouncing the complicity of these companies in bringing 'blood coltan' to the mobile-phone industry by sending text messages to a large number of mobile-phone users. Boston-based Cabot Corporation, the world's second-largest processor of coltan, also declared – albeit in the midst of glutted market – that it would not buy any more coltan from the Great Lakes region.[18] The UK government attempted but initially failed to pressure Premier Oil out of Burma, while it succeeded in preventing the London Stock Exchange from listing Oryx, a diamond company operating in the DRC, in which the Zimbabwean military has interests.[19] The Canadian government, by contrast, took a 'conciliatory and positive dialogue' path towards the government of Sudan despite an incriminating report commissioned by the Canadian Ministry of Foreign Affairs asserting the complicity of Canadian oil company Talisman in human-rights abuses in southern Sudan.[20]

The appeal of sanctions and market-access restrictions, however, should not gloss over their negative impacts. Following an extensive study, the DRC-based Pole Institute argued that, despite the clear links between war and coltan, and the need to stop the illicit economy organised and controlled by armed groups, 'the people of the Kivu would not gain [from an embargo], but lose one of their very few remaining sources of income'.[21] Rather than imposing market restrictions, however, an ideal regulatory framework should address the dilemma of sustaining the local economy in the interest of the population while weeding out activities supporting belligerents. At the same time, 'buying out' belligerents through economic incentives may be a necessary step towards achieving peace. This challenge thus entails both economic supervision and wealth sharing.

Supervising economies and sharing wealth

As military demobilisation and electoral monitoring accompany most peace processes, so a war economy needs to be 'demobilised' and 'monitored' in order to help avoid the resumption of conflict. This economic aspect of a peace process is generally neglected and too often placed under the initiative of belligerents jockeying for key economic positions within the new authority or simply embezzling funds to re-arm. Beyond sanctions and global regulatory measures, practical regulatory frameworks can be set up to deprive belligerents of revenues they could use to follow a double agenda of peace transition and rearmament, as happened repeatedly in Angola, Cambodia, Colombia, Liberia, Sierra Leone and Sri Lanka. Internationally supervised tax collection and budgetary allocation using escrow funds could seek to ensure that populations and public institutions benefit from resource revenues (see Figure 1). Direct payment of resource revenues could be made to the population, as suggested in the case of Iraq. This would have the advantage of clearly distributing a 'peace dividend' to the most needy, and partly address the problem of lack of representation and accountability through broad-based taxation affecting many resource-dependent countries. Businesses themselves would be deterred from operating outside the scheme through a system of incentives, such as secure legal ownership, and deterrents, such as effective sanctions. If successful, and in the absence of alternative sources of support, opting out of a peace process would become prohibitively costly for belligerents. Like all instruments of control, the effectiveness of such scheme would depend in part upon the characteristics of the targeted resource sector and the economic incentives attached. Alluvial diamonds, for example, would likely remain a resource difficult to control.

The UN Oil-for-Food Programme, set up to limit the humanitarian impact of UN sanctions against Iraq, represented an early example of such scheme. The clear failures of this programme – relating in part to the lack of oversight by UNSC members over its financial aspects rather than its weapons and security implications – stand as a warning for future such schemes. The UN expert panel on Liberia has recommended a similar programme to prevent the proceeds of the Liberian shipping and corporate registry[22] from financing arms-sanctions-busting. The Security Council took a step in that direction by calling upon the government of Liberia to establish transparent and internationally verifiable audit regimes over its use of revenues derived from both its shipping and corporate registry as well as the timber industry. These audits were required to demonstrate that the revenues were not used for busting sanctions, but for 'legitimate social, humanitarian and development purposes'.[23]

The World Bank has indirect oversight over the Chad–Cameroon Petroleum Development and Pipeline Project, which is intended to forestall conflict and prioritise the allocation of oil revenues to social sectors. After nearly three decades of civil war, negotiations in the mid-1990s between the northern-dominated government and the main southern rebellion made it feasible for oil companies to develop fields in southern Chad. The consortium of oil companies viewed the World Bank as the 'centrepiece of its risk

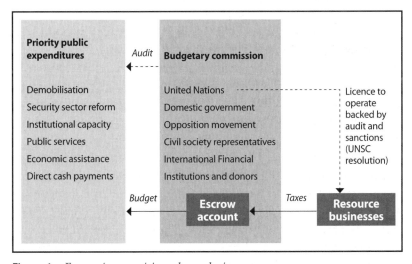

Figure 1 – *Economic supervision scheme during peace processes*

reduction strategy': as well as providing funds for the pipeline project, the Bank assisted the Chad government with revenue management and the implementation of social and environmental programmes.[24] As 'moral guarantor' of the scheme, the World Bank established an International Advisory Group to observe the implementation of the project and make recommendations to both governments. To manage the estimated $1.5bn in forecasted revenue over the next 28 years, the Chad parliament placed this revenue into an offshore escrow account, allocated it to social and environmental priority sectors, and submitted it to a public auditing and an oversight committee. However, President Idriss Deby, who came to power through a military coup in 1990, used $4m from the oil development signature bonus to purchase weapons; this move triggered an outcry from human-rights and environmental NGOs, and led the World Bank and the IMF to threaten the government with exclusion from their debt-relief programme.

As part of an on-going peace process, in January 2004, the government of Sudan and the Sudanese People's Liberation Army (SPLA) signed a wealth-sharing agreement over oil resources, dividing oil revenues between the two parties. The scheme, however, is not independently supervised. A similar wealth-sharing scheme had been established as part of initiative previous the 1997 Sudan–Khartoum Peace Agreement signed between the Sudanese government and local southern armed factions in control of most of the oil area; this allowed the government to attract major international investments and companies, and oil was flowing by 1999. The resulting peace, however, was short lived. Not only was the agreement boycotted by the SPLA, but some of the southern factions felt 'cheated' by the government and rejoined the SPLA. Military hostilities in the oil area sharply increased as the SPLA sought to bolster its negotiating position by intensified attacks against oil installations and army units, and the government conducted devastating counter-insurgency attacks on local populations. Although such a scenario is less likely following the 2005 agreement, given the participation of the SPLA this time around, some SPLA officials had suggested that without proper external supervision, each party is likely simply to increase its military capacity, resulting in an intensification of the war.[25] Iraq will most likely be in need of some form of independent mechanism overseeing the workings of the oil sector, although the future Iraqi government is likely to resist any such supervision.

In the long term, the key to having resources serve to consolidate peace rather than as fuel for war, however, is strong democratic control over resource revenues rather than a weak external mode of regulation.[26] The latter is a risky strategy: by providing a façade of legitimacy through partial control of resource rents by a few selected civil-society representatives and foreign advisors, an external supervision scheme could serve to dampen pressure for democratisation and thus open the way for renewed conflict.

Leveraging international aid

Aid conditionality can help 'strengthen incentives for ending conflict and discourage a return to war'.[27] The effectiveness of aid conditionality – conditions set by donors upon recipients for aid provision – on regulating resources flows is highly dependent upon the importance of this aid to the targeted recipient, which would most frequently be a state assisting a rebel group with resource trafficking.

Aid- or trade-dependent countries such as Thailand, and, to a lesser degree, Liberia, have responded to aid conditionality. By the mid-1990s, the US government was under growing pressure, most notably from Global Witness, to subject the Thai government to aid conditionality, so that it would end its assistance to the Khmer Rouge. The US Congress included to this end specific clauses in the Foreign Operations Act, first threatening to end military assistance in 1996 and then all assistance in 1997. Congress further asked for a report on this matter from the secretary of state. Further, the IMF pressured Thailand to end its acquiescence to unauthorised exports by cancelling part of its Enhanced Structural Adjustment Fund loan in late 1996. Such pressures significantly contributed to the demise of the Khmer Rouge movement. In 2000, following years of 'quiet diplomacy' and aid commitment towards the Liberian government to improve human rights and regional stability, the EU suspended about US$50m in aid on evidence of government involvement in arms and diamond trading with the RUF. This suspension had reportedly little effect on the trade itself but was part and parcel of broader measures that constrained the RUF's main ally and promoted an end to the conflict.[28]

Increasing corporate responsibility

Businesses, from petty gems traders to oil majors, often decline to play a direct role in setting the context of, or prolonging armed conflicts.[29] The business perspective is often that conflict-affected countries are 'intrinsically unstable areas': as US Vice-President Dick Cheney, a former oil-company executive, noted, 'the problem is that the good Lord didn't see fit to put oil and gas reserves where there are democratic governments'.[30] Most companies only acknowledge influence over their direct activities and areas of operations, and emphasise the positive economic impact of their presence. Although conflict-prone countries should not face further economic marginalisation, resource businesses need to recognise and help address potential negative consequences that their activities may have, including:[31]

- exacerbating inequalities, greed and grievances by aggravating inequalities or increasing economic rents amenable to factional control;
- sustaining poor governance by paying taxes to unrepresentative and repressive authorities or participating in corruption;
- impinging upon human rights, in relation to local livelihoods, labour, or the use of force to protect corporate interests;
- bankrolling belligerents and hindering peace processes.

The responsibility of companies in hostilities varies, from being simple intermediaries in supply chains to exerting complex forms of influence, including diplomatic and military support. Rarely do belligerents operate resource exploitation schemes on their own, and all require business intermediaries – from local 'barefoot entrepreneurs' to international brokers and multinationals – to access commodity, financial or arms markets. While some businesses simply attempt to cope with a deteriorating political and security context, while others see the possibility of competitive advantage. As oil expert Thomas Waelde observes from the history of the oil industry, 'at the beginning of most corporate or individual successes ... was usually some bold, rarely very ethical, exploitation of commercial opportunities blocked to competitors by politics'.[32] Oil businesses gain a competitive advantage by engaging with 'pariah states' that other companies avoid to protect their good name or to preclude sanctions. Following years of opposition by human-rights campaigners and pressure from officials in their home governments, British company Premier Oil sold its Burmese assets in 2002 and Canadian company Talisman Energy sold its Sudanese operations in 2003. In both cases, buyers were Asian companies facing less pressure from campaigning groups. Unocal, a US energy company present in both Afghanistan and Burma, ended its pipeline project through Afghanistan in 1998 as the US militarily retaliated against al-Qaeda's terrorist attack that year in Nairobi.

Countries in conflict also constitute a valuable 'niche market' for businesses whose competitive advantage lies in their risk-taking mentality, political savvy, or connections with security services. At best, 'pioneers' of the international economy provide local jobs, humanitarian assistance and tax revenue much needed for social services. At worst, opportunistic 'bottom-feeders' directly support war criminals in their financial and arms dealings. Often, there is a thin line between these two categories. Junior companies in particular seek out markets characterised by high political risks or legal barriers such as sanctions, creaming off otherwise easily accessible resource reserves or preparing the ground for investment by larger businesses.[33] To access and secure resources in these unpopular places, businesses often associate themselves with dubious brokers or private military corporations. In other cases, businesses deal with arms traders paid through natural resource concessions or mortgaged resource production, or directly deal in arms.

Business interests also 'invest' in rebel factions with an eye to accessing resource areas in the near term while paying the government to keep longer-term options open, or vice-versa. During the Algerian war of independence (1954–62), the Italian oil company ENI reportedly supplied money and arms to the FLN in return for future 'considerations'.[34] Western businessmen did the same with UNITA leader Savimbi in Angola during the 1980s. From late 1996, many foreign companies supported the Rwandan- and Ugandan-backed AFDL as it gained control of eastern and southern Zaire (renamed DRC after Kabila's May 1997 takeover), including key mining sites. With its huge demand for arms and wealth of mineral resources, Angola became a prime target of businesses juggling political relations, arms dealing and natural-resources brokering in the 1990s. Some companies may even appear politically progressive as a result, seeking contacts with separatist groups whose aspirations have international legitimacy. In West Papua/Irian Jaya, currently under contested Indonesian rule, corporate interests may have enticed the Indonesian government to consider further political and fiscal devolution for the province in order to defuse tension.[35]

Two important first steps towards reducing political conflict and violence over resource exploitation sites would be to improve communications among different stakeholders (mainly companies, local residents and governmental authorities) and to establish security measures respectful of human rights. Some conflicts are aggravated by the overestimation of revenues flowing from resource exploitation projects. A 'myth of plenty' can be reinforced by the culture of secrecy among managing governmental institutions and companies. Conflicts are more easily avoided if realistic revenue forecasts, emphasising the finite nature of resource reserves and vagaries of international prices, along with the costs and benefits of resource exploitation, are communicated to the population and host authorities from the outset of the project.

Often taking place in remote areas with politically marginalised populations, resource exploitation sites can easily become the private domain of businesses and governments. The multiplicity of ownership claims over land-based resources often increases the potential for conflict, but participatory management and a fair allocation of revenues or compensation can attenuate this tendency. Numerous forms of co-ownership and management have emerged, such as long-term concessions and production-sharing agreements between the state and the private sector, or community-based natural resource management schemes providing local populations some degree of control over resource access.[36] Internal tensions within local communities also demand constant attention, communication and conflict management skills on the part of authorities and businesses.

There is a case for the engagement of business in conflict prevention, along the following principles:

- strategic commitment on the part of the management, translated through explicit policies on human rights, corruption and security;
- risk and impact assessment of the company's core business and social investment activities;
- dialogue and consultation with key stakeholder groups on a regular basis;
- partnership and collective action with other companies, government and civil society organisations to address sensitive issues such as fiscal accountability and human-rights records, and to invest in practical projects;
- evaluation and accountability through performance indicators, independent verification and public reporting. [37]

The deployment and use of security forces to prevent banditry, enforce ownership rights, and protect staff and infrastructures against violence is a particularly sensitive issue. At the initiative of the US and UK governments, in partnership with a small number of major companies and NGOs, a set of guidelines, 'Voluntary Principles on Security and Human Rights', were established to guide companies in their operations. These principles, adopted by participating companies and NGOs in December 2000, relate to risk assessment, and interactions between companies and public or private security forces. There are practical barriers to their viability. Firstly, the application of the principles has produced delays and difficulties in security practices. Secondly, the agreement does not carry the force of law or even mandate compliance assessment by the US and UK governments.

Like most voluntary codes of conduct, the efficacy of the Principles on Security and Human Rights pivots on the good will and capacity of shareholders and individual managers. The profit motive and competitive pressures constitute formidable countervailing forces. Because many corporations still see a systematic trade-off between shareholder value and social responsibility, more progressive businesses are wary of the advantages that could accrue to less socially minded competitors. Strong compliance with standards or binding legal regulation can also carry their own negative externalities. Although major transnational corporations can set positive standards within the sectors they dominate and act as 'gatekeepers' against bad practices, these corporations can also abuse this situation to marginalise smaller competitors and emerging markets. For example, the blanket labelling of all African artisanal diamonds as 'conflict diamonds' makes diamonds coming from industrial mines and developed countries such as Australia and Canada more marketable as 'clean diamonds' and potentially impedes the development of third-world enterprises. There is thus a need to balance global legal standards for business practices, entrepreneurship in developing countries and foreign investment incentives in the developing world. In fact, the move to regulate the global diamonds trade is being accompanied by both international aid and business partnerships to assist authorities and small businesses in producing countries to adapt to the new regulatory environment (see below).

Prosecuting business accomplices

The judicial prosecution of sanctions busters and commercial actors engaging with war criminals is still in its infancy. International judicial institutions such as the International Criminal Court have, for the most part, abandoned the idea of prosecuting the business accomplices of war criminals, in part due to corporate pressure. Basic legal and administrative requirements for implementation by member states have been recently promoted through the Interlaken Process on 'smart sanctions', but enforcement by national authorities remains a crucial loophole in the UN sanctions regime. Human-rights activists and victims of human-rights abuses are increasingly using extraterritorial jurisdiction over resource businesses involved in human-rights violations, provided, for example, by the US by the Alien Tort Claims Act. The Bush administration, however, has repeatedly voiced its opposition to such use of the legislation, with the Justice Department filing a brief arguing that the use of ATCA 'bears serious implications for our current war against terrorism, and permits claims to be easily asserted against our allies in [the] war [on terror]'.[38] The most significant formal measure could be the International Convention for the Suppression of the Financing of Terrorism, which theoretically applies to businesses dealing financially with terrorists and war criminals. This entered into force on 10 April 2002, following UNSC Resolution 1373 urging member states to ratify it in the light of the 11 September terrorist attacks on the US.[39] This convention could be applied to diamond dealers linked to al-Qaeda financing networks.

The UN Convention against Transnational Organised Crime, with its focus on the criminalisation of the laundering of proceeds of crime (Article 6) and corruption (Article 8), potentially covers the laundering of natural resources obtained through criminal offences, defined either domestically (such as armed rebellion or 'grand corruption') or internationally (such as war crimes). There is, however, no specific reference to this type of criminal activity or to natural resources as potential proceeds within the text of the convention. The convention could be complemented by a protocol specifically addressing the Illicit Exploitation of and Trafficking in Natural Resources. Modelled upon the UN Protocol against the Illicit Manufacturing of and Trafficking in Firearms, Their Parts and Components and Ammunition, such a legal instrument could define and criminalise a broad range of offences relevant to conflict-prevention and conflict-termination purposes, including financial complicity in war crimes, application of certification and sanctions regimes, and the legality of resource exploitation. It could also make mandatory a number of useful measures, such as record-keeping and public access to import/export and revenue figures, and the financial transparency of governments and resource businesses.

Regulating conflict diamonds

In 1978, the head of the de Beers diamonds cartel, Harry Oppenheimer, could think 'of no commodity less susceptible to dangers from UN sanctions than diamonds'. [40] Operating in South Africa, where the apartheid regime had just come under a Security Council arms embargo, and in South African-ruled Namibia, the threat of commodity export sanctions that would affect de Beers's interests seemed real. Yet tens of millions of dollars could be smuggled out from these countries in an attaché case, and buyers could be found on the international market. Ironically, 20 years later, de Beers supported UN sanctions against 'unethical' diamonds – not because they bankrolled apartheid or colonial regimes, but because they were 'conflict diamonds ... used by rebel movements or their allies to finance conflict aimed at undermining legitimate governments'.[41] De Beers had become the target of campaign by Global Witness in 1998 criticising the role of its open-market purchasing practices in financing UNITA through the purchase of UNITA diamonds on the world market. A year later, the company pulled out of Angola's open diamonds market as a broader NGO campaign argued that 'most people would be horrified to learn that their diamond jewellery had financed the purchase of landmines or guns in one of Africa's brutal conflicts'.[42] This decision to pull out was taken in the context of de Beers' testy relations with the Angolan government and a new marketing strategy in which the cartel would reduce its huge price-regulating stocks and move from being 'buyer of last resort' to 'supplier of choice' for the high end of the market.[43]

Although the role of diamonds in fuelling several African conflicts has long been known, international attempts to regulate this 'conflict resource' were slow to materialise. Early attempts, such as the proposal of a ban on imports of UNITA diamonds by Belgian MPs in 1993, had failed for several reasons, including the ease with which diamonds could be concealed, the difficulty of identifying rough diamonds in mixed shipments, and the customary tolerance of the industry and importing countries for illicit diamonds smuggled to avoid taxation in producing countries.[44] Politically, the delay in imposing UN sanctions was explained by the UN's obligation 'to behave impartially between the two sides in helping them to implement the [Lusaka] Protocol … [even if] there were doubts about the sincerity of Savimbi's commitment to do so'.[45] There may have also been some resistance by pro-UNITA elements within Western governments and political elites who had close links with the mining industry.[46] Furthermore, it is possible that business interests interfered, as several mining multinationals were eager to develop large-scale mining operations in alluvial fields and probably believed they could only do so successfully with the approval of UNITA, or in the absence of a civil war. Finally, importing countries such as Belgium, where about 80% of rough diamonds are internationally traded, were wary that drastic regulations on traders could drive the industry to a less regulated venue elsewhere.

It took UN exasperation at the lack of UNITA cooperation and the failure of successive UN missions in Angola to bring about the first sanctions on UNITA diamonds in 1997. Sanctions regimes on Angola, Sierra Leone and Liberia were useful for promoting the disengagement of the most visible mining or trading companies and imposing greater difficulties of access to legitimate markets. Beyond their financial impact, sanctions also served to politically marginalise rebel groups and their allies, essentially by portraying them as 'greed-driven' bandits or 'spoilers'. The denunciation of major sanctions-busters, including African heads of states, by UN expert panels was relatively effective in undermining far-reaching support, including diplomatic assistance and arms trafficking. The imposition of sanctions on Liberia in 2001 following recommendations by the UN expert panel also gave credibility to this new policy of sanction implementation. Yet not only was the international community initially slow to react to a problem long identified, but since sanctions came into place, only a handful of suspected violators have been arrested on charges of tax evasion, arms trafficking, and false papers – but not sanction-busting. In this respect, military interventions by national armies and mercenaries were arguably more effective than regulatory measures in controlling rebel access to diamond revenues; however, such military successes have often proved short lived if not followed by effective conflict resolution and demobilisation.

In support of UN sanctions, campaigns against conflict diamonds led by Global Witness and Partnership Africa Canada raised the profile of the issue to an international level and initiated significant industry reform. After the diamond industry's initial denial of the problem and its argument that the provenance of diamonds could not be determined, the industry recognised its vulnerability to a consumer boycott in light of the discretionary luxury status of diamonds and their image-driven value.[47] The fear of a consumer boycott, as experienced by the fur industry in the 1980s, led the diamond industry as well as producing and key trading countries to react by organising public-relations events and international negotiations to set up the Kimberley Process Certification Scheme.

The Kimberley Process Certification Scheme

Launched in May 2000 in Kimberley, South Africa's first diamond town, by African producing countries eager to protect their trade, the Kimberley Process consisted of a dozen international meetings, drawing together government officials from up to 38 different countries as well as representatives of the diamonds industry and NGOs. The dominant position of De Beers in the marketing chain and the implementation of the country-specific certification scheme in Sierra Leone facilitated the establishment of this process, as did the existence of the Belgian Diamond High Council (HRD) in the main trading centre of Antwerp, and the creation during the 2001 diamond industry conference of a World Diamonds Council, which was specifically tasked by member industries with eradicating the trade in conflict diamonds. The positive, if sometimes tense, engagement between NGOs, government and industry was relatively innovative and succeeded in preserving the interests of most of the industry and an economic sector that employs several hundred thousand people and is critical to the economy of Botswana and Namibia. The Kimberley Process Certification Scheme (KPCS) established a voluntary system of industry self-regulation centred on a certification process requiring all participating countries:

- not to trade in rough diamonds with any non-participant;
- to ensure that each export shipment of rough diamonds is accompanied by a certificate and to require such certificate on all imports;
- to establish a system of internal controls on exploitation and trade, as well as a mode of shipment eliminating conflict diamonds from any exports;
- to collect and maintain data on production, import and export statistics;
- to cooperate with other participants, providing assistance in fulfilling the minimal requirements, and to maintain transparency, including through external reviews.[48]

The KPCS was concluded in late 2002 after two years of negotiations. By then the conflicts in both Angola and Sierra Leone had been officially declared over. The first major decision taken through the KPCS was the exclusion of the Republic of Congo (Brazzaville) from the list of participants following the improper issuance of Kimberley Process certificates to launder diamonds smuggled from neighbouring countries. The scheme plays a significant role in preventing diamonds from fuelling war and in limiting corruption and the politicisation of this sector by bringing about greater transparency and curbing illicit trading. The scheme remains under criticism for using peer-review mechanisms rather than a more independent type of monitoring, and for failing to directly regulate individual companies, leaving this task to national authorities. Finally, there is still a need to regulate non-participating countries that are likely to continue smuggling illicit and conflict diamonds; also, to help set up diamond polishing and cutting activities to short-circuit the rough diamonds markets and directly link the supply chain to jewellers operating outside the KPCS.

National legislation and control schemes
Parallel to the Kimberley Process Certification Scheme and UN Security Council resolutions, both exporting and importing countries have taken legal and institutional initiatives. Among exporting countries, Sierra Leone was the first to set up, in 2000, with the assistance of the Belgian HRD, a chain of custody linking individual mines to the diamond buying market in Antwerp following UN Security Council Resolution 1306. Diamond mines and local buyers and exporters are licensed and monitored by the government, and an independent diamond valuer, who assesses their value, certifies that they do not come from rebel-controlled areas and examines legal diamond shipments. Certified diamonds are then taxed and shipped in

sealed containers along with a certificate of origin, while identification details of the parcels are send to Antwerp for checking upon arrival.

Despite receiving the approval of the UN sanctions committee, the Sierra Leone regime has several loopholes. Firstly, the scheme is essentially geared to increasing government diamond revenues; while low, the export tax of 3% induces many traders to smuggle their best diamonds out of the country, thereby sustaining illicit networks that channel conflict diamonds. Secondly, monitoring is open to bribery since officials are paid only about $50 per month and have no logistical support, often making them reliant upon the very miners and traders they have to monitor. Thirdly, some licensed mines bordering RUF territory were poorly monitored and alleg-edly used as 'reception' or laundering centres for RUF diamonds.[49] Finally, while sanctions on diamonds were adopted against Liberia, illicit trade through Gambia continued to provide a conduit for illicit diamonds until the KPCS started.[50] Legal diamond exports rose from $1.5m in 1999 to over $70m in 2003, out of a production estimated at about $250m.

In contrast to Sierra Leone, Angola has adopted a marketing monop-oly to facilitate the control of diamond trade. The scheme has met some success, but it is relatively non-transparent and raises fears that it privately benefits members of the presidential entourage. As the monopoly imposed low prices, it also resulted in the smuggling of the best stones to the DRC, a country whose monopoly system had rapidly collapsed under the pres-sure of entrepreneurs accustomed to avoiding taxation from a predatory state that would give little in return.[51]

Among importing countries, the US represents 65% of the world's diamond jewellery market, and legislators led by Tony Hall pushed the Clean Diamonds Trade Act through Congress. Passed by the House of Representatives in late 2001, the Act defines safeguards against the trade in conflict diamonds by exporting countries, provides the president with authority to block rough diamond imports from non-implementing coun-tries, permits US Customs agents to seize suspect diamonds or jewellery and authorises $10m per annum to assist countries in implementing the Kimberley Process Certification Scheme.

The international regulation of diamonds through a combination of Security Council sanctions and the Kimberley Certification Process should nevertheless provide, if not a complete solution, at least a signif-icant means of limiting the role of diamonds in fuelling war and bring much needed transparency to an industry in which opacity sustains illicit commerce, money-laundering and corruption. This system of regula-tion could provide a model for other 'conflict resources' – in particular, timber, which is probably the most prevalent yet least regulated resource involved in armed conflicts. The criminalisation of sanctions-busting, and the promotion of international investigative and judicial work occurring as a result of the regulation of conflict commodities, should help tackle the role of resource businesses in enabling conflict.

The example of conflict diamonds can inform policy reforms and initiatives on the role of resources in fuelling war. Besides promoting a comprehensive legislation on resource governance and conflict commodities, a general sequence of measures for the regulation of conflict resources based on available instruments should include:

- early-warning systems identifying the existence of conflict resources, based on media, business, and NGO reports, possibly by a permanent UN monitoring mechanism;
- a UN Security Council resolution setting up expert panels and considering sanctions on conflict resources pending adequate national certification or international supervision of the resource sector;
- assessment of the potential role of military intervention and UN peacekeeping deployment in targeted resource areas;
- international review of legislation and practices in resource extraction and trading in neighbouring countries and other areas identified as transit or importing countries;
- criminal prosecution of sanction-busters using national and international legislation;
- aid and assistance to targeted countries to facilitate compliance with international certification schemes and Security Council resolutions.

Beyond initiatives curtailing the laundering of conflict resources through the international market, there is a need to ascertain that revenues generated by natural resources reach government budgets, to be used fairly and efficiently through strong and legitimate institutions. To facilitate this process, a global regulatory regime for key resources, such as oil, minerals and timber, is needed to promote transparency and accountability in resource and financial flows. The Extractive Industries Transparency Initiative initiated by the British government deserves much attention and support in this regard. Finally, resource-dependent countries would benefit from an international economic and aid environment that prevents economic shocks and maximises resource revenues and economic diversification.

CONCLUSION

The geopolitics of natural resources has long been a strategic concern for both exporting and importing states. Western powers' concerns over 'resource wars' have been largely palliated by the end of the Cold War and greater flexibility of international trade. Continued supply dependence, rising demand for raw materials, and recent armed confrontations and instability in key areas such as the Persian Gulf continue to place resources high up on the geopolitical agenda. This study has presented reasons behind the links between resources and armed conflicts – some of which remain hotly debated – and possible solutions. The cases examined here suggest that improving resource governance should be a strategic goal for the international security community.

The fact is that many conflicts are largely funded by resource revenues. During the 1990s, accessible and internationally marketable resources financed no less than 20 major conflicts. This is not to argue that those wars were only financed or motivated by the control of resources, but that resources figured prominently on the agendas of belligerents. Given the concentration of wars in poor countries with few foreign-earning sources, resources are likely to remain the economic focus of most belligerents in years to come. Even if conflict resources come under greater regulatory pressure, it is likely that criminal networks and unscrupulous businesses will pursue trading, especially those already involved in arms trafficking.

In addressing this issue more effectively, the main priorities are greater awareness and tighter controls on resource trade, earlier and stronger imposition of targeted sanctions, and peacekeeping mandates allowing for the military capture and supervision of resource production sites. International instruments to prevent or terminate conflicts financed by natural resource exploitation would move from 'shaming' international actors to formalising punishments and sanctions against individuals as well as corporations. During peace processes, the international community should follow a principle of 'economic demobilisation' limiting the risk of renewed conflict and building the resource-governance capacity of new institutions.

Conflicts around resource exploitation deserve broader attention than that of resource companies and local resource-management agencies. Economic, environmental and socio-cultural issues present numerous challenges to local populations, business interests, the state, and global environmental and human rights networks.[1] Organised opposition to processes of globalisation unaccountable to local interests and growing demand for raw materials will only increase such adversarial politics and the need for more effective dialogue. While most conflicts are can be peacefully negotiated, social protest movements and small-scale skirmishes can polarise local communities and result in full-scale civil wars. Central and regional authorities, as well as international agencies, should monitor and engage more actively with such conflicts to address problems early and prevent escalation.

Finally, long-term stability in resource-dependent regions will depend on their developmental outcomes. There is a clear pattern of economic underperformance and governance failure among resource-dependent countries. Given continued resource dependence by many developing countries, translating resource exploitation into political stability and economic development will remain a central issue on the development and security agenda in the years to come, particularly for Africa and the Middle East. The stability of many autocratic regimes relying on resource rents has proven to be illusory, when not outright dangerous, as in the case of Iraq.

A clear priority is to link resource exploitation and institutional capacity building more systematically. Such linkages should ensure that revenues first serve the basic needs and security of local populations, thereby reinforcing the stability and legitimacy of state authorities. International institutions, along with resource companies, should step in to ensure greater transparency, accountability and fairness in the allocation of resource revenues when domestic governments fail to do so. The voluntary Extractive Industries Transparency Initiative and the Chad–Cameroon Petroleum Development and Pipeline Project offer valuable templates in this regard. Direct revenue disbursement to local citizens should be considered. International regulatory approaches to corporate social responsibility are key. Beyond resource companies, the participation of financial operators needs to be reinforced, notably in terms of freezing and repatriating embezzled resource revenues – as illustrated by the case of Swiss government repatriation of Nigerian funds in 2005. Producing countries and international organisations should consider how revitalised commodity agreements and complementary financial mechanisms – such as commodity price insurance schemes – could maximise and buffer resource revenues. The issue of illegal resource exploitation, most notably that of timber, requires specific attention. The problem of debt in resource-dependent countries should be addressed specifically, through debt relief and compensatory financial mechanisms. New international institutions may be required to balance the diverse interests of exporting and importing countries, for example, in the oil sector.

Reforming resource governance to ensure better developmental and security outcomes is a massive challenge. As the cases of Iraq and Nigeria make clear, however, the consequences of resource dependence on developmental outcomes and security cannot be ignored. Transparency, accountability and fairness in resource governance will not come out of individual initiatives by corporations necessarily driven by profits, governments protecting their interests behind the veil of sovereignty, or civil-society and international organisations lacking leverage. As the case of conflict diamonds has made clear, only concerted efforts between industry, governments and civil society can bring about significant changes. For the resource curse to turn into a blessing, such coalitions need to strategically orient resource governance towards the needs and aspirations of populations in resource-exporting countries.

Acknowledgements

The author gratefully acknowledges support and comments from Richard Auty, Karen Bakker, Mats Berdal, Paul Collier, Charmian Gooch, Terry Karl, Morlai Kamara, David Keen, Eric Leinberger, Mike Moore, Gilberto Neto, Jenny Pearce, Michael Ross and Jonathan Stevenson.

Notes

Introduction

1   On the case of diamonds in West Africa, see Richard Snyder and Ravi Bhavnani, 'Diamonds, Blood, and Taxes: A Revenue-Centered Framework for Explaining Political Order', *Journal of Conflict Resolution*, vol. 49, no. 4, August 2005 (forthcoming).

2   Barbara Crossette, 'U.N. Chief Faults Reluctance of U.S. to Help in Africa', *New York Times*, 13 May 2000.

3   Michael T. Klare, *Resource Wars: The Changing Landscape of Global Conflict* (New York: Henry Holt, 2001), p. 213.

4   Paul Richards, 'Are "Forest Wars" in Africa Resource Conflicts? The Case of Sierra Leone', in Nancy L. Peluso and Michael Watts (ed.), *Violent Environments* (Ithaca, NY: Cornell University Press, 2001), p. 65; Kenneth Roth, 'Precious Stones Don't Kill, Guns Do: Enforce Arms Embargoes', *Los Angeles Times*, 21 July 2000.

5   Raphael Marques, cited in Paul Salopek, 'CEOs of War Bleed Angola', *Chicago Tribune*, 2 April 2000.

6   Richard M. Auty, *Sustaining Development in Mineral Economies: the Resource Curse Thesis* (London: Routledge, 1993); Michael L. Ross, 'The Political Economy of the Resource Curse', *World Politics*, vol. 51, no. 2, January 1999, pp. 297–322.

Chapter One

1   Jeffrey D. Sachs and Andrew M. Warner, 'Natural Resources and Economic Development. The Curse of Natural Resources', *European Economic Review*, vol. 45, 2001, pp. 827–38; Richard M. Auty (ed.), *Resource Abundance and Economic Development* (Oxford: Oxford University Press, 2001).

2   Michael L. Ross, *Extractive Sectors and the Poor*, Oxfam America Report (New York: Oxfam, 2001), p.16.

3   About 20% of the population receives 59% of the national income and annual diamond public revenues account for about $1,000 per capita. See Ralph Hazleton, *Diamonds Forever, or Diamonds for Good? The Economic Impact of Diamonds in Southern Africa* (Ottawa: Partnership Africa Canada, 2002), pp. 3–8.

4   Carlos Leite and Jens Weidmann, 'Does Mother Nature Corrupt? Natural Resources, Corruption, and Economic Growth', *IMF Working Paper WP/99/85* (Washington DC: IMF, 1999); Philippe Le Billon, 'Fuelling War or Buying Peace: The Role of Corruption in Armed Conflicts', *Journal of International Development*, vol. 15, no. 4, 2003, pp. 413–26.

5   John Mason, 'Abacha to Return $1bn in Funds to Nigeria', *Financial Times*, 17 April 2002.

6   Michael L. Ross, 'Does Oil Hinder Democracy?', *World Politics*, vol. 53, April 2001, pp. 325–41; Terry L. Karl, *The Paradox of Plenty: Oil Booms, Venezuela, and other Petro-states* (Berkeley, CA: University of California Press, 1997).

7   Paul Collier and Anke Hoeffler, *Greed and Grievance in Civil War* (Washington DC: World Bank, October 2001). This data set includes 52 wars taking place between 1960 and 1999.

8   Indra de Soysa, 'The Resource Curse: Are Civil Wars Driven by Rapacity or Paucity?', in Mats Berdal and David Malone, *Greed and Grievance: Economic Agendas in Civil Wars* (Boulder, CO: Lynne Rienner, 2000), pp. 113–35. The data set includes conflict over 25 battle-deaths between 1989 and 1998.

9   Michael L. Ross, 'What Do We Know About Natural Resources and Civil War', *Journal of Peace Research*, vol. 41, no. 3, 2004, pp. 337–56.

10 See for example Graham A. Davis, 'Learning to Love the Dutch Disease: Evidence from the Mineral Economies', *World Development*, vol. 23, no. 10, 1995, pp. 1765–79. On average fuel producers economically outperformed non-mineral economies, but this historical analysis of economic success among mineral exporters however is only based on two years (1970 and 1991) and exclude several countries affected by conflicts.

11 Richard Auty, 'Industrial Policy Reform in Six Newly Industrializing Countries: The Resource Curse Thesis', *World Development*, vol. 22, no.1, January 1994, p. 12.

12 Charles Tripp, *A History of Iraq* (Cambridge: Cambridge University Press, 2000), p. 281.

13 Abbas Alnaswari, *The Economy of Iraq: Oil, Wars, Destruction of Development and Prospects, 1950–2010* (Westport, CT: Greenwood Press, 1994); Jahangir Amuzegar, *Managing the Oil Wealth: OPEC's Windfalls and Pitfalls* (London: Tauris, 1999).

14 Financial obligations include debt ($127bn), war reparations ($199bn) and pending contracts ($57bn). This amount does not include about $100bn war compensation claims by Iran. See Frederick D. Barton and Bathsheba N. Crocker, *A Wiser Peace: An Action Strategy for a Post-Conflict Iraq* (Washington DC: Center for Strategic and International Studies, January 2003).

15 Statement by Dr Isam al Khafaji Before the US Senate Foreign Relations Committee, 24 September 2003, p. 2.

16 *Iraq Backgrounder: What Lies Beneath*, Middle East Report 6 (Brussels: International Crisis Group, 1 October 2002).

[17] The Oil-for-Food Programme itself generated $65bn dollars, of which $4.4bn would represent in illicit surcharges on oil sales and after-sales charges on suppliers. About $5.7bn worth of oil was also smuggled out of Iraq. See 'Recovering Iraq's Assets: Preliminary Observations on U.S Efforts and Challenges', GAO-04-579T, General Accounting Office, 18 March 2004; C Rosett, 'The Oil-for-Food Scam: What Did Kofi Annan Know, and When Did He Know It?', *Commentary Magazine*, 16 April 2004.

[18] Oil revenue was $26bn in current dollars in 1980, equivalent to about $52bn in 2004, while the population has nearly doubled from 13 million in 1980 to 25m in 2004. Ahmed Jiyad, 'The Economic Debt Trap: Iraq's Debt, 1980–2020', *Arab Studies Quarterly*, vol. 23, no. 4, 2001, pp. 15–58.

[19] Relatively similar economic situations prevail for most of the population of West African oil producing countries, see Ian Garry and Terry Karl, *Bottom of the Barrel: Africa's Oil Boom and the Poor* (Baltimore, MD: Catholic Relief Service, June 2003). The main difference with African countries is that Iraqis did enjoy a higher level of welfare prior to the mid-1980s.

[20] Michael D. Shafer, *Winners and Losers: How Sectors Shape the Development Prospects of States* (Ithaca, NY: Cornell University Press, 1994); Karl, *The Paradox of Plenty*, p. 15.

[21] Jan Dehn, 'The Effects on Growth of Commodity Price Uncertainty and Shocks', mimeo, Working Paper 2455 (Washington DC: World Bank, 2000); Michael Renner, *The Anatomy of Resource Wars* (Washington DC: Worldwatch Institute, 2002).

[22] Paul Cashin, C. John McDermott and Alasdair Scott, 'Booms and Slumps in World Commodity Prices', *Reserve Bank of New Zealand Discussion Paper No. G99/8*, December 1999.

[23] Osmel Manzano and Roberto Rigobon, 'Resource Curse or Debt Overhang?', *NBER working paper no. w8390*, July 2001.

[24] Crude oil US 1st purchase price (wellhead) in 1996 $, www.wtrg.com.

[25] Net hydrocarbon revenue per capita had fallen from a peak of $613 in 1981 to $172 in 1988, see Ali Aïssaoui, *Algeria. The Political Economy of Oil and Gas* (Oxford: Oxford University Press, 2001), pp. 8–9, 15.

[26] Martine-Renée Galloy and Marc-Éric Gruénais, 'Fighting for Power in the Congo', *Le Monde Diplomatique*, November 1997.

[27] William Reno, *Corruption and State Politics in Sierra Leone* (Cambridge: Cambridge University Press, 1995).

[28] Peter Uvin, 'Rwanda: the Social Roots of Genocide', in E. Wayne Nafziger, Frances Stewart and Raimo Vayrynen (eds), *Weak States and Vulnerable Economies: Humanitarian Emergencies in Developing Countries* vol. 2 (Oxford: Oxford University Press, 2000), pp. 175–6.

[29] Kevin M. Murphy, Andrei Shleifer and Robert W. Vishny, 'The Allocation of Talent: Implications for Growth', *The Quarterly Journal of Economics*, vol. 106, May 1991, pp. 503–30.

[30] Mick Moore, 'Political Underdevelopment: What Causes Bad Governance?', *Public Management Review*, vol. 3, no. 3, 2001, p. 389.

[31] Ross, 'Does Oil Hinder Democracy', pp. 356–7.

[32] Philip Lane and Aaron Tornell, 'The Voracity Effect', *American Economic Review*, vol. 89, March 1999, pp. 22–46.

[33] Michael L. Ross, *Timber Booms and Institutional Breakdown in Southeast Asia* (Cambridge: Cambridge University Press, 2001).

[34] Ross, *Extractive Sectors and the Poor*, p. 15.

[35] *Times* (London), 21 November 1996, p. 31, cited in Kenneth A. Rodman, *Sanctions Beyond Borders: Multinational Corporations and U.S. Economic Statecraft* (Lanham, MD: Rowman & Littlefield, 2001).

[36] Jean-François Médard, 'Oil and War: Elf and "Françafrique" in the Gulf of Guinea', mimeo, 2002.

[37] Håvard Hegre, Tanja Ellingsen, Scott G. Gates and Nils Petter Gleditsch, 'Toward a Democratic Civil Peace? Democracy,

Political Change and Civil War, 1816–1992', *American Journal of Political Science Review,* vol. 95, no. 1, 2001, pp. 33–48.

[38] Xavier Sala-i-Martin and Arvind Subramanian, 'Addressing the Natural Resource Curse: An Illustration from Nigeria', WP/03/139 (Washington DC: International Monetary Fund, 2003).

[39] Michael Watts, 'Antinomies of Community: Some Thoughts on Geography, Resources and Empire', *Transactions of the Institute of British Geographers,* vol. 29, no. 2, 2004, p. 204.

[40] John Connell, 'The Panguna Mine Impact', in Peter Polomka (ed.), 'Bougainville Perspectives on a Crisis', *Canberra Papers on Strategy and Defence,* no. 66, p. 43.

[41] Charlie Pye-Smith and Bernice Lee, 'Armed Conflict and Natural Resources: The Case of the Minerals Sector', workshop report, MMSD/IISS, London, 11 July 2001, p. 4.

[42] *Sudan: The Human Price of Oil,* Report no. AFR 54/01/00 (Washington DC: Amnesty International, 2000).

[43] Hugues Leclerc, 'Le rôle economique du diamant dans le conflict Congolais', in Laurent Monnier, Bogumil Jewsiewicki, and Gauthier de Villers (eds), *Chasse au Diamant au Congo/Zaire* (Paris: L'Harmattan, 2001), p. 60.

[44] Michael Watts, 'Petro-Violence: Community, Extraction, and Political Ecology of a Mythic Commodity', in Nancy L. Peluso and Michael Watts (eds), *Violent Environments* (Ithaca, NY: Cornell University Press, 2001); pp. 189–212.

[45] James C. Scott, *Weapons of the Weak: Everyday Forms of Peasant Resistance* (New Haven, CT: Yale University Press).

[46] Neil Harvey, *The Chiapas Rebellion: The Struggle for Land and Democracy* (Durham, NC: Duke University Press, 1998).

Chapter Two

[1] David Keen, *The Economic Functions of Violence in Civil Wars,* Adelphi Paper 320 (Oxford: Oxford University Press for the IISS, 1998), p. 11.

[2] Philippe Le Billon, 'Angola's Political Economy of War: The Role of Oil and Diamonds, 1975–2000', *African Affairs,* vol. 100, 2001, p. 67.

[3] Interview with Kennedy Hamutenya, Ministry of Mines and Energy, Ottawa, April 2002.

[4] Global Witness, *A Rough Trade: The Role of Companies and Governments in the Angolan Conflict* (London: Global Witness, 1998).

[5] Approximate price in producing country during the 1990s, adapted from Richard Auty, 'Natural Resources and Civil Strife: A Two-Stage Process', *Geopolitics,* vol. 9, no. 1, 2004; and interview with Gavin Hayman, Global Witness, London, June 2002.

[6] Auty, *Resource Abundance.*

[7] Douglas Farah, *Blood from Stone: The Secret Financial Network of Terror* (New York: Broadway, 2004).

[8] Interviews with Congolese diamond buyers, Malange/Angola, 2001; 'Violations of Security Council Sanctions Against UNITA', S/2000/203, p. 27; Filip De Boeck, 'Garimpeiro Worlds: Digging, Dying & "Hunting" for Diamonds in Angola', *Review of African Political Economy,* vol. 28, no. 90, 2002, pp. 554–5.

[9] Françoise Chipaux, 'Des mines d'emeraude pour financer la résistance du Commandant Massoud', *Le Monde,* 17 July 1999.

[10] Global Witness, *Corruption, War and Forest Policy. The Unsustainable Exploitation of Cambodia's Forests* (London: Global Witness, 1996).

[11] Philippe Le Billon, 'The Political Ecology of Transition in Cambodia 1989–1999: War, Peace and Forest Exploitation', *Development and Change,* vol. 31, no.4, 2000, pp. 785–805.

[12] Report of the Panel of Experts on the Illegal Exploitation of Natural Resources and Other Forms of Wealth of the Democratic

Republic of the Congo (S/2001/357), 12 April 2001, UN Secretariat, New York.

13 Marc-Antoine de Montclos, 'Libéria: des prédateurs aux "Ramasseurs de Miettes"', in François Jean and Jean-Christophe Rufin (ed.), *Economie des Guerres Civiles* (Paris: Hachette, 1996), p. 281.

14 'Petroleum Companies Hire Army to Curb Terrorism', *Crime and Justice: The Americas*, vol. 9, no. 5, 1996.

15 Interview with Jenny Pearce, Bradford University, June 2002.

16 Interview with James Fennel, ArmorGroup, London, January 2002.

17 Interview with Luc Van Zandvliet, Collaborative for Development Action, June 2002.

18 Roland Pourtier, '1997: Les raisons d'une guerre "incivile"', *Afrique Contemporaine*, no. 186, 1998, pp. 7–32; interview with Pascal Lissouba, London, January 2002. Controlling the north of the country, Nguesso could also have benefited from logging revenues of timber exports via Gabon and Cameroon (*La lettre du Continent*, 13 January 2000).

19 Pourtier, 'Les raisons d'une guerre "incivile"', p. 7.

20 Nkossa was the name of an oil field recently awarded to French oil company Elf Aquitaine. Rémy Bazenguissa-Ganga, 'Les milices politiques dans les affrontements', *Afrique Contemporaine*, no. 186, 1998, p. 52.

21 Stephen Ellis, *The Mask of Anarchy. The Destruction of Liberia and the Religious Dimension of an African Civil War* (London: Hurst, 1999), p. 108.

22 Marites Danguilan Vitug, *The Politics of Logging. Power from the Forest* (Philippines: Philippine Center for Investigative Journalism, 1993.

23 Michael L. Ross, 'How Does Natural Resource Wealth Influence Civil War? Evidence from 13 Cases', *International Organization*, vol. 58, no. 1, 2004, pp. 35–7.

24 Tony Hodges, *Western Sahara: The Roots of a Desert War* (Westport, CN: Lawrence Hill, 1983), p. vii.

25 Nazaruddin Sjamsuddin, 'Issues and Politics of Regionalism in Indonesia: Evaluating the Acehnese Experience', in Lim Joo-Jock and Vani (eds), *Armed Separatism in Southeast Asia* (Singapore: Institute of Southeast Asian Studies, 1984).

26 Cited in Peter Polomka (ed.), 'Bougainville Perspectives on a Crisis', *Canberra Papers on Strategy and Defence*, no. 66, p. 8. See also Volker Boge, 'Mining, Environmental Degradation and War: The Bougainville Case', in Mohamed Suliman (ed.), *Ecology, Politics and Violent Conflict* (London: Zed Books, 1999), pp. 211–28; Karl Claxton, 'Bougainville 1988–98', *Canberra Papers on Strategy and Defence*, no. 130, 1998.

27 Cited in Sharon E. Hutchinson, *Nuer Dilemmas: Coping with Money, War, and the State* (Berkeley, CA: University of California Press, 1996), p. 9.

28 Interview with SPLA officials, Nairobi, November 2001.

29 Lynn Horton, *Peasants in Arms: War and Peace in the Mountains of Nicaragua, 1979–1994* (Athens, OH: Ohio Center for International Studies, 1998), p. 155.

30 Alain Labrousse, 'Colombie-Pérou: Violence Politique et Logique Criminelle', in Jean and Rufin, *Economies des Guerres Civiles*, p. 386.

31 Report of the Panel of Experts on the Illegal Exploitation of Natural Resources.

32 Final Report of the Judicial Commission of Inquiry into Allegations into Illegal Exploitation of Natural resources and Other Forms of Wealth in the Democratic Republic of Congo 2001 (May 2001–November 2002), Kampala, November 2002.

33 Thomas Pakenham, *The Boer War* (London: Weidenfeld and Nicolson, [1979] 1997).

34 *Algeria. The Political Economy of Oil and Gas* (Oxford: Oxford University Press, 2001), pp. 47–9.

35 Alexander A. Arbatov, 'Oil as a Factor in Strategic Policy and Action: Past and Present', in Arthur H. Westing (ed), *Global Resources and International Conflict:*

Environmental Factors in Strategy Policy and Action (Oxford: Oxford University Press, 1986), p. 34.

[36] Jean-Marie Balancie and Arnaud de La Grange, *Mondes Rebelles. Guerres Civiles et Violences Politiques* (Paris: Michalon, 1999), pp. 446–8.

[37] Interview with Professor Séverin Mugangu, Université Catholique de Bukavu, April 2002.

[38] Interview with Terry Karl, Stanford University, June 2002.

[39] Cited in Nate Thayer, 'Rubies Are Rouge, Khmer War Effort Financed by Gem Finds', *Far Eastern Economic Review*, 7 February 1991, pp. 29–30.

[40] Nazih Richani, *Systems of Violence: The Political Economy of War and Peace in Colombia* (Albany, NY: SUNY, 2002).

[41] No correlation is found by Paul Collier, Anke Hoeffler and Mans Söderbom, 'On the Duration of Civil War', *World Bank Policy Working Paper 2861*, 2001; but most of the case study literature argues the contrary, see Ross, 'How Does Natural Resource Wealth Influence Civil War?, pp. 35–67.

[42] Stephen John Stedman, Donald Rothchild and Elizabeth Cousens (eds), *Ending Civil Wars: The Implementation of Peace Agreements* (Boulder, CO: Lynne Rienner, 2002) find no peace agreement successfully implemented between 1987 and 2000 in places with valuable and easily marketable commodities; Michael Doyle and Nicholas Sambanis, 'International Peacebuilding: A Theoretical and Quantitative Analysis', *American Political Science Review*, vol. 94, no. 4, 2000, pp. 779–801, find primary commodity exports negatively correlated with peace building successes between 1945 and 1997.

[43] Interview with Roland Marchal, Centre d'études et de recherches internationales (CERI), Paris, April 2002.

[44] United Nations, Report of the Panel of Experts on Violations of Security Council Sanctions Against UNITA, S/2000/203 (New York: United Nations Secretariat, 2000).

[45] Assis Malaquias, 'Diamonds Are a Guerilla's Best Friend: the Impact of Illicit Wealth on Insurgency Strategy', *Third World Quarterly*, vol. 22, no. 3, 2001, pp. 311–25.

[46] On 'bottom up' violence, see Keen, *The Economic Functions of Violence in Civil Wars*.

[47] Interview with the author, Cambodia, January 2001.

[48] Interviews with former KR soldiers and commanders, Pailin and Along Veng, 2001.

Chapter Three

[1] Christopher J. Gilbert, 'International Commodity Agreements: an Obituary Notice', *World Development*, vol. 24, 1996, pp. 1–19.

[2] Interview with Andrea Mogni, EU Commission, Brussels, November 2001.

[3] For short- to medium-term horizons, derivative markets – in the form of futures, options, or swaps – can hedge commodity price risk. Yet these are complex and mostly accessible by international buyers, banks and brokers, rather than small producers, and do not cover long-term risks. 'New Options for the Poor?', *The Economist*, 19–25 August 2000.

[4] Africa–Caribbean–Pacific–European Community Partnership Agreement signed in Cotonou, Benin, 23 June 2000.

[5] 'The President's Comprehensive Corporate Reform Agenda', White House Press Release, 9 July 2002.

[6] Thomas I. Palley, 'Combating the Natural Resource Curse with Citizen Revenue Distribution Funds: Oil and the Case of Iraq', *Foreign Policy in Focus* (Silver City, NM: Interhemispheric Resource Center, December 2003); Martin and Subramanian, 'Addressing the Natural Resource Curse'. On direct cash disbursements, see J. Hanlon, 'It is Possible to Just Give Money

to the Poor', *Development and Change*, vol. 35, no. 2, 2004, pp. 375–83.

7   Paul Toyne, Cliona O'Brien and Rod Nelson, *The Timber Footprint of the G8 and China: Making the Case for Green Procurement by Government* (Gland: WWF International, 2002).

8   Based on a survey of 22 armed conflicts involving resources between 1989 and 2004, see Philippe Le Billon, *Natural Resources and the Termination of Armed Conflicts: Share, Sanction, or Conquer?* unpublished manuscript, January 2005.

9   Security Council resolution S/RES/1295 (2000), para. 27.

10 Charles Recknagel, 'Oil Smuggling Produces High Profits', RFE/EL, 21 June 2000.

11 Steven L. Myers, 'UN Concludes, Fining Shell, That Tanker Carried Iraq Oil', *New York Times*, 26 April 2000.

12 Interview with Ambassador Mahmoud Kassem, Chairman of DRC expert panel, New York, December 2001.

13 Interview with Ian Smillie, Ottawa, March 2002.

14 Gregory Mthembu-Salter, *An Assessment of Sanctions Against Burundi* (London: Action Aid, 1999), pp. 18–19.

15 Executive Order No. 13067, http://www.ustreas.gov/ofac/t11sudan.pdf; S.180 EAH, House of Representatives, 15 November 2001.

16 Community Aid Abroad, 'Submission to the Joint Parliamentary Committee on Corporations and Securities inquiry into the Corporate Code of Conduct Bill 2000', December 2000.

17 http://www.eitransparency.org/

18 Blain Harden, 'A Black Mud From Africa Helps Power the New Economy', *New York Times*, 12 August 2001.

19 Interview with former Oryx employee, London, November 2000.

20 John Harker, 'Human Security in Sudan: the Report of a Canadian Assessment Mission (Ottawa: Canadian Ministry of Foreign Affairs, 2000).

21 *The Coltan Phenomenon* (Goma: POLE, 2001), p. 4.

22 Many companies register themselves in Liberia or other so-called 'offshore' territories – to benefit from corporate legislation and low-level taxes; however, they do pay some minimal taxes, such as a registration fee, hence the 'corporate registry'.

23 UN Security Council resolution S/RES/1408 (2002), art. 10.

24 Korinna Horta, *Questions Concerning The World Bank and Chad/Cameroon Oil and Pipeline Project* (New York: Environmental Defense Fund, March 1997).

25 Interview with SPLA official, Nairobi, November 2001.

26 Interview with Charmian Gooch, Global Witness, London, June 2002.

27 James Boyce, *Investing in Peace: Aid and Conditionality after Civil Wars*, Adelphi Paper no. 351 (Oxford: Oxford University Press for the International Institute for Strategic Studies, 2002), p. 73.

28 Interview with Alex Vines, UN expert panel member, London, February 2001.

29 John Bray, *No Hiding Place: Business and the Politics of Pressure* (London: Control Risk Group, 1997).

30 Dick Cheney was chairman of oil servicing company Halliburton, cited in *Petroleum Finance Week*, 1 April 1996. Ironically, this argument would suggest that the US, and the state of Texas in particular, are undemocratic.

31 See Jane Nelson, *The Business of Peace: The Private Sector as a Partner in Conflict Prevention and Resolution* (London: The Prince of Wales Business Leaders Forum, 2000); Damien Lilly and Philippe Le Billon, *Regulating Businesses During Armed Conflicts* (London: Overseas Development Institute, 2002).

32 Thomas Waelde, 'Legal Boundaries for Extraterritorial Ambitions', in John V. Mitchell (ed.), *Companies in a World of Conflict: NGOs, Sanctions and Corporate Responsibility* (London: Royal Institute of International Affairs and Earthscan, 1998), p. 178. For a story of the oil industry, see Daniel Yergin, *The Prize: The Epic Quest for*

Oil, Money and Power (New York: Simon and Schuster, 1992).

33 See Pierre Baracyetse, 'L'enjeu geopolitique des Sociétés Minières Internationales en République Démocratique du Congo (ex-Zaire)', mimeo, December 1999.

34 Cited in Ali Aïssaoui, *Algeria. The Political Economy of Oil and Gas* (Oxford: Oxford University Press, 2001), p. 49.

35 Interview with Oxford Papuan Rights Campaign, 2001.

36 Daniel Buckles (ed.), *Cultivating Peace: Conflict and Collaboration in Natural Resource Management* (Ottawa: IDRC, Washington DC: World Bank, 1999).

37 Nelson, *The Business of Peace: The Private Sector as a Partner in Conflict Prevention and Resolution*, p. 1.

38 Jim Lobe, 'Attorney-General Attacks Key Law', Inter Press Service, 15 May 2003

39 See www.un.org/law/cod/finterr.htm. War criminals are included in the broad definition of the original convention but the EU, among others, greatly narrowed it; interview with Anthonius de Vrie, EU commission, Ottawa, March 2002.

40 Edward Jay Epstein, *The Rise and Fall of Diamonds* (New York: Simon and Schuster, 1982).

41 For a full definition, see Kimberley Process Working Document no. 1/2002, Ottawa, 20 March 2002. The definition is problematic insofar as it does not define 'legitimate governments'.

42 'Campaign Launched To Stop Billion Dollar Diamond Trade From Funding Conflict In Africa', Fatal Transactions Campaign, 3 October 1999.

43 Francesco Guerrera and Andrew Parker, 'The Changing Face of the Diamond Industry', *Financial Times*, 11 July 2000, p. 16.

44 François Misser and Olivier Vallée, *Les Gemmocraties: L'Economie Politique du Diamant Africain* (Paris: Desclée de Brouwer, 1997).

45 Interview, Sir Marrack Goulding, former UK Ambassador to Angola and former head of the UN Department for Peace Keeping Operations, Oxford, October 1999.

46 Neil Cooper, 'Conflict Goods: The Challenges for Peacekeeping and Conflict Prevention', *International Peacekeeping*, vol. 8, no. 3, 2001, p. 28.

47 Interviews with De Beers officials and HRD, London and Antwerp, 1998 and 2001. On identification, see *Possibilities for Identification, Certification and Control of Diamonds* (London: Global Witness, 2000).

48 Adapted from the Kimberley Process Working Document no. 1/2002.

49 Interviews with residents and diamonds traders near Punduru, Sierra Leone, April 2001.

50 Interview with Alimami Wurie, director of Ministry of Mines, Freetown, April 2001.

51 Christian Dietrich, *Hard Currency: The Criminalized Diamond Economy of the Democratic Republic of Congo and Its Neighbours*, Occasional Paper no. 4 (Ottawa: Partnership Africa Canada, 2002), p. 35.

Conclusion

1 Al Gedicks, *Resource Rebels: Native Challenges to Mining and Oil Corporations* (Cambridge, MA: South End Press, 2001).